Luiz G. Spoladore
Specialist in University Education Didactics
Master degree in Medical Sciences (UFF)
Former University Assistant Professor (FUSSV)
From the Science and Technology Department (MS)

New Atomic Theory

A proposal

Electronic furnace
Atomic capacitor
Genetic pole
Alpha-appendix
Phylogenetic code
Electron-projectile
Magnetic tunnel
Alpha decay
Band theory of solids
Radioactive families
Crookes' tube
Atoms and Deuterotomes
Lenton and Quantum
Neutron captures
Gamma rays

With 129 original illustrations (photos and drawings)

Rio de Janeiro – RJ
Brasil
2007

Translated by N. M. Pereira (Technical Translator)

CIP – BRASIL. CATALOGAÇÃO NA FONTE
SINDICATO NACIONAL DOS EDITORES DE LIVROS, RJ

s749n

Spoladore, Luiz G. (Luiz Galileu), 1935-
New Atomic Theory: a proposal
/ Luiz G. Spoladore; [translated by Luiz G. Spoladore],
Rio de Janeiro: Argônio, 2007
218p.: ill;

Translation: Nova Teoria Atômica
Includes bibliography
ISBN 978-85-60599-01-1

1. Atomic theory. 2.Nuclear physics.
I. Title.

07-2983 CDD 539.7
CDU 539.1

CIP – Brazil Cataloging in the Source
Sindicato Nacional dos Editores de Livros, Rio, RJ – Brasil
Spoladore, Luiz G.
New Atomic Theory (Original title: Nova Teoria Atômica), with addition
Includes Indeces
1. Atomic Theory. 2.Nuclear physics. I. Title
s749n 07-2983 CDD 539.7 CDU 539.1

Argônio Editora

Argônio – Editora e Distribuidora de Livros Ltda
e **Livraria Argônio**
Rua da Lapa, 120 – loja 101 (térreo) – Centro
CEP 20021-180 - Rio de Janeiro – RJ – Brasil

Fax: +5521 2221-4527
E-mail: riobarao@yahoo.com.br e livrariaargonio@ig.com.br
Website: http://www.livrariaargonio.hpg.com.br
Printed in Rio – Brasil
2007

Index

To the master of Physics and Chemistry,

Prof. Dr. J. J. Puls,

with special reverence

Foreword

No area in science has its own way; in other words, random knowledge is wishful thinking. Though a great number of areas are closely intertwined, that does not necessarily mean that a lead-in is available towards comprehension. Merging contents of two or more areas usually fosters a more complex or state-of-the-art wisdom. This paper combines what is already known about industrial and atomic capacitors, unveiling thus new insights into Nuclear Physics.

The contents in this book are limited to those theories that attempt to reshape the patterns related to the atoms of hydrogen and, in so doing, giving rise to the deuterotomic theory, both based on the Atomic Capacitor Theory. At the very moment, we have not been able to further delve into references as to the physics of particles, save for electrically negative leptons and neutrinos in addition to proton and neutron baryons, undoubtedly, the main particles existing in the atomic theory. Neither the mesons π^o, K±, K^o, η, η', nor the baryons Σ+, Σ–, Δ, Ω– etc. have been referred to elsewhere in the New Atomic Theory. It is common knowledge, however, that they are formed by a number of fundamental particles known as quarks identified by the letters: **u** (up), **d** (down), **s** (sideway or strange), **c** (charmed quark) and **b** (bottom), besides others from recent findings. Three quark particles make up a meson or a baryon in triad chains, such as: sss, uds, uus etc. Mesons are unstable particles with a zero spin while quarks have a ½ spin.

The New Atomic Theory put forward here is analytical, interpretative and, above all, comparative. It is anything but experimental and its contents aim simply at casting some light on this field. For the time being, the theory could not be developed any further and can therefore be regarded as a fait accompli; in other words, there is no room for the theory to be further upgraded or downgraded. The lentons theory, though, seems to be the weak part in this paper, but in no way could it be left out for it was a cornerstone for the New Theory to be developed in keeping with the current quantum theory. The initial hypotheses on the lentonian structure of the so-called electrically-negative lepton and the proton and neutron baryons rendered it possible to understand the emission of photons (quantum) by electrons and gamma rays by neutrons in transition to proton, due to the loss of nucleus excitation following the emission of beta particles by an intranuclear neutron. The lentonian theory highlights the origin of X-ray photons (from electrons hitting a target wall) and gamma ray photons (from beta-emitting neutrons of intranuclear location). The lentons (or quanta) that lie interplate in electrons and protons alike are ejected from the interior of the particles through the action of a shock or a state of excitation (in the characteristic X-ray emission) in the specific case of electrons, or through a nuclear excitation, as is the case of protons (that took on that status after beta emission by a neutron had occurred).

The process of calculating the capacitance of the atoms and deuterotomes mentioned in this book are also introductory in view of their high complexity, the types of arrangements that atomic capacitors are subject to, the variable number of plates in dielectrics located between the protons and the variations in thickness **d** found in the capacitance equation, owing to the entrance or exit of a neutron.

Broadly speaking, the contents of a technical paper and how its subjects are unfolded, are fit to meet the requirements of a target audience. In this book, however, this target audience turned out to be particularly hard to define. It was then taken into account that any technical paper attempting to reformulate the Atomic Theory in force would be of interest to people already knowledgeable in this field, whatever their cultural level or professional background. One can easily realize that after a century of an Atomic Theory based on the model proposed by Rutherford-Bohr and Thomson, practically taught in every school and thoroughly adopted by most leading brains–an inseparable part of a century-old world culture–the proposal of a New Atomic Theory is likely to come up against fierce opposition. No less strong an argument in favor of the current Atomic Theory is the technology available that has led to breathtaking monitored atomic explosions by nuclear fission and thermonuclear fusion, to unique scientific discoveries in the area of electronics (such as television, computer, cellular communication, etc.), besides the technological application of radioactive properties (thermoelectric power stations, radiotherapy of tumors, medical examinations by scintilography, etc.), to mention but a few.

A Theory can remain unchanged for a century or longer than that. Nevertheless, should it contain gray areas, as is the case of the nucleus of atoms, whose composition the current Atomic Theory fails to provide a well-founded account, save for, perhaps, the Theory of magic numbers proposed by Meyer, it is, most likely, in want of reformulation. It is certainly stunning to envision an atom (or deuterotome) with an "electronic crown" located internally, as proposed by the "electronic furnace" in heavy and radioactive deuterotomes like uranium; this idea, though, is by far scary. Only a line of logical reasoning, based on the assumption that deuterium can be an atomic capacitor, could serve as a fundamental basis to attempt to reformulate the Atomic Theory in force. The tracks of thought in this line of reasoning are not entirely self-explanatory. The first steps taken failed to produce conclusive evidence; consequently, new doors had to be sought after until those areas of knowledge could pinpoint their individual interdependence. Having this in mind, a goal was finally reached, thus making it possible to mentally travel through the deuterian structures that make up the elements ranging from helium to uranium, besides what one could learn from the physical volume-reduction phenomenon in noble gases, the nature of the genetic and chemical poles in elements, the phylogenetic code of heavy deuterotomes belonging to radioactive families, the radioactivity mechanisms, besides other less important points.

Accordingly, we are absolutely convinced that this proposal to reformulate the Atomic Theory in force will draw the attention of a diversified yet qualified and attentive audience–the aim of this New Atomic Theory.

The Author. Rio, December 8, 1998

E-mail: riobarao@yahoo.com.br **Website**: http://www.livrariaargonio.hpg.com.br

CHAPTER 1

The origin of matter

1.1 – Lenton and Lentonian space

Space and time are fundamental greatnesses in Physics. Space is filled with primary units deprived of mass and electric load. These units can be regarded as single flakes set apart from each other. Each primary unit may be termed lenton (L. lentus = calm) and the space then occupied by lentonian space. A lenton juxtaposed over another creates multilaminate plates. When lentons cluster round at a certain point, others are fully emptied. The void space that results from the concentration of lentons brings about the first pressure difference inside the lentonian space. The appearance of this new greatness fuels the movement of lentons as well as the formation of new lenton plates. The plates formed by the aggregation of lentons produce energy of electric nature in their interior and the unit thus formed makes up the electron-lenton unit, or simply put, the electron.

The aggregations of lentons produce electrons and protons alike; in order for this aggregation to occur no external energy is necessary, for lentons are potentially energized even though devoid of electric load. Therefore, electrons and protons are composed of lenton plates, which leave empty spaces between the plates. Both are capable of capturing free lentons in the interstices of their plates. That reception is greater in protons, but the lentons lying in the electron interstices are more subject to being released or emitted upon the collisions they are liable to or due to the variations occurring in their energy levels. The two following schematic figures represent an electron and a proton, both formed by lenton plates with free lentons in their interstices.

Upon entering the physical space of another, a lenton creates a new energetic and pulsating unit with an internal movement for accommodating its structure. This new structure becomes stable and capable enough of incorporating other lentonian units. This dual pulsating and vibrating feature progressively leads the electron to acquire kinetic energy.

Lentons are basic units that fill up space and provided with aggregating properties among themselves. Consequently, two units of lentons aggregate have been formed, namely the **electron** with x.1 lenton plates and the **proton** with x.1836 lenton plates where the value of **x** is unknown.

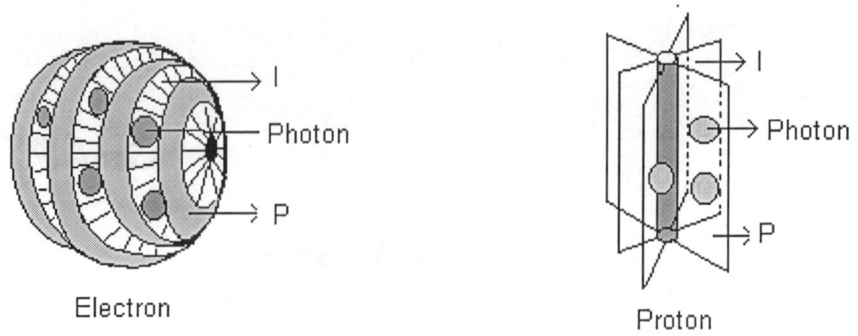

Fig 1. Electron composed of x.1 lenton plates with free lentons (photons) in the interstices of the plates (p). Proton composed of x.1,836 lenton plates with free lentons (photons) in the interstices.

The displacement of lentons to make up electrons and protons in the lentonian space, thus creating empty spaces and pressure differences, entices other lentons to keep on moving, and this displacement becomes progressively faster. Thus, the synthesis of electrons and protons from lentons gets quicker. Lentons are clustered in shells roaming slowly in the lentonian space–a shell slides over the other. As a result of these movements, lentons keep constant touch with each other and because of their combining properties they produce laminate plates of lentons. These plates then capture other small units, become electrically loaded and generate the basic electric unit of matter, the electron or the larger unit, the proton, which is a reservoir for lentons. Upon the formation of electron and proton, the first electric tension breaks out outright between two structures, for both can store energy. Thus, the limit of matter is not the atom itself, or its particles, but rather the lenton. The lenton is the smallest unit of energy and matter and has a null mass. An electron e_1 upon striking one other electron e_2 loses lentons through ejection; part of them can be hurled into the other electron (in the interstitial space between the lentons plates) and the remaining part is thrown outside as photonic waves. The concepts on lenton and quantum bear a close resemblance when one considers a free lenton provided with kinetic energy for the latter can also act as a photon. The following figure shows the shock mechanism between two electrons and the loss of lentons towards the environment and the transfer of lentons to the electron with less kinetic energy.

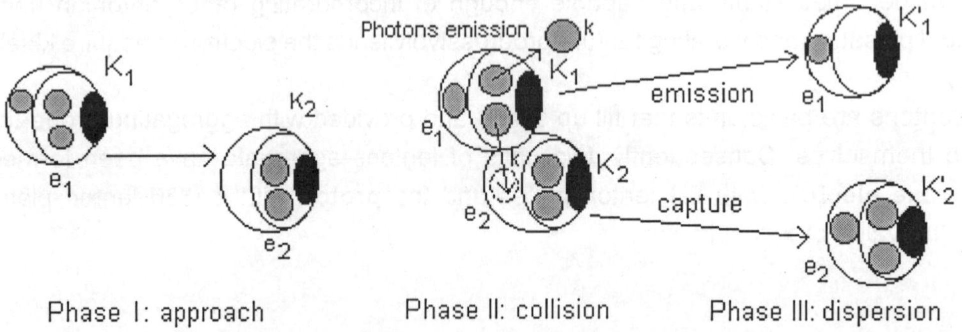

Fig. 2. Scheme showing electron e_1 provided with kinetic energy K_1 striking electron e_2 with less kinetic energy, K_2. As a result of the shock, two lentons (photons) k are ejected from electron e_1: 1) one moves outwards as photonic wave; 2) the other moves towards the interstice of electron e_2, whose kinetic energy rises up to K_2'.

Given that lentons are real sources of energy, this means that they keep a close relationship with the quantum energy and can be analyzed in keeping with the Quantum Theory. Thus, it can be demonstrated that $K_1 > K_2$ and $K_2' > K_1'$, the Law of Conservation of Energy is upheld as per the following equation:

$$K_1 + K_2 = K_1' + K_2' + 2k$$

where K_1 is the kinetic energy of electron e1 that emitted two photons k (one to the exterior and the other to the interior of electron e2); K_2 is the kinetic energy of electron e_2 that has captured a photon k. Thus, $K1' = K_1 - 2k$ and $K2' = K_2 + k$. Also: $K_1 > K_2 > K_1'$. Due to the emission of two photons of energy k from K_1, the final kinetic energy of electron e_1 is K_1', hence, less than that of K_1.

Deuterotomes whose electrons can be stimulated by incoming energy are capable of emitting or capturing photons in the same fashion. The Quantum Theory demonstrates that electrons can go through a condition of greater energy by the absorption of a **v** frequency photon, on condition that the difference between the energy of final (f) and initial states (i) be the same as the photon energy (hv). The following equation points out the energy conservation:

$$Ef - Ei = hv$$

This is the energy conservation equation where **h** is Planck's constant and **v** the frequency. The electron can also shift from a condition of higher energy to another of lower level of energy through the emission of photons, often given by **v** in the equation $\mathbf{v = \Delta E/h}$, where $\Delta E = Ei - Ef$ or $\Delta E = hv$, ΔE being the energy difference between the two levels. The following illustration shows an electron with initial energy **Ei** at a higher orbital zone, which acquires a final energy **Ef** when going down towards a lower zone after emitting a photon with energy (**E**).

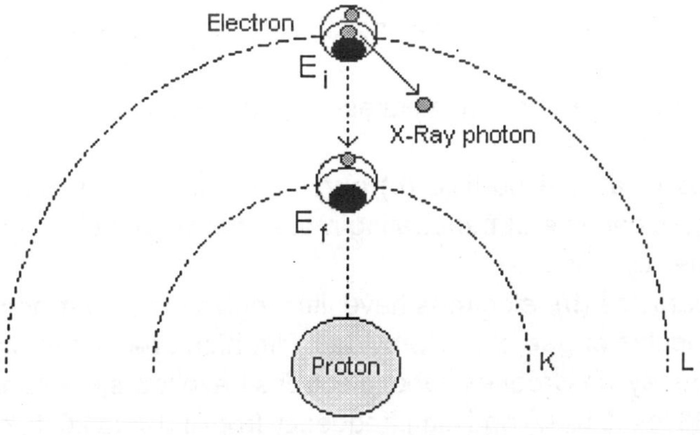

Fig. 3. Illustration of a deuterotome D with proton (P) and electron (e), the latter situated at the orbital zone of higher energy level with initial energy Ei, moving towards

the shell of Lower energy closer to proton (P) with final energy Ef after emitting a photon of energy (E).

1.2 – Electron, proton and neutron

a) - Electrons: Electrons consist of a small number of lenton plates juxtaposed over each other with free lentons in their interstices. They are units of little weight in comparison with protons. If the electron has x.1 lentons plate, then the proton has x.1836 plates. Thus, an electron weighs 1,836 times less than a proton. The two units are composed of energy. The energetic mass found in an electron is similar to that contained in a proton, that is, one electron-volt. Due to the proton's physical greatness and its laminate structure similar to that of an electron, free lentons can be among its plates. Thus, the energy level contained in a proton is considerably different from that of an electron. In light of the potential difference, one can say that the proton has positive load (higher) whereas the electron has negative load (lower). This electron can remain in an orbital movement round the proton. The distance between the proton and the orbital electron will depend on the load of lentons captured by the lentonian plates of the protons in their interstices. This unit makes up the atom of light hydrogen (H1/1). The electron mass can vary in line with its speed. At a speed of 100,000 km/s its mass is equal to 1/ 1,836 of the proton, but at 270,000 km/s this mass is reduced by half (1/3,700).

Depending on its physical condition, the electron has different functions:

1) - It neutralizes the electric load of the proton of light hydrogen (H1/1), during electron-conversion in neutron which will become the dielectric of the atomic capacitor,

2) - It neutralizes the electric load in the pair of proton reciprocally (resulting from the capture of alpha particles at the synthesis stage of matter),

3) - It controls the maximum number of free valences in deuterotomes, mainly those with a great number of deuterons in their structure, for instance bromine (Br35), arsenic (As33), tellurium (Te52) etc., and leaves free only the last seven valences. Otherwise, that is, should neutralization fail to occur, deuterotomes would be found with up to 31 valences available, if one takes into account that the series of P-shell electrons has up to 31 deuteriums after noble gas xenon (Xe54) and that the elements in the N-shell and O-shell electrons could have up to 17 valences. This fails to occur as a result of the said function,

4) – It functions as one of the two plates–the negative one–that make up the atomic capacitor,

5) – It consists of a beta particle (b) pursuant to the radioactivity phenomenon of heavy deuterotomes capable of penetrating matter between alpha particles and gamma rays, entering water by 1 to 2 cm,

6) – As beta particles (b), electrons have little ionization power and each particle can ionize 5 to 7 molecules of gas, being provided with high speed and short wavelength,

7) – As cathode rays in Crookes' tube, electrons have low speed and long wavelength ranging from 0.793 to 0.0189 Å and high frequency from 1.10^6 to 20.10^6 cycles per second (hertz).

Electrons can at least have two physical states: an **electric state** and a **magnetic state** and are interchangeable. Depending on the condition they undergo, the electron can be pleomorphic and shift from a state to the other. While in electric state they possess the following fundamental properties: high density, low plasticity, small diameter, small volume, identifiable electric load, impenetrability and can be stored in capacitors. If in magnetic state, they have opposite fundamental properties, namely: low density, high plasticity, large diameter, great volume, undetectable electric load, penetrability, and cannot be captured by capacitors such as the electric-electron due to the properties referred to above. The following scheme points out all these properties:

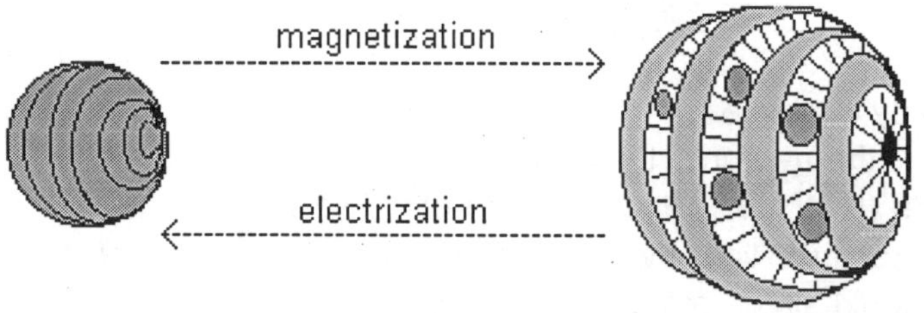

Electron in electric state

1–high density
2–low plasticity
3–small diameter
4–small volume
5–identifiable electric load
6–can be stored in a capacitor
7–high-speed

Electron in magnetic state

1–low density
2–high plasticity
3–large diameter
4–great volume
5–undetectable electric load
6–cannot be stored in a capacitor
7–slow and induced displacement

Fig. 4 - Fundamental properties of the electron's physical states.

b) Neutrons: Neutrons stem from an atom of light hydrogen, that is, they consist of a proton and an electron (H1/1). Under this form, light hydrogen does not participate in the composition of deuterotomes. Neutrons are necessary to make up the atoms of deuterium and tritium. Without them, the synthesis would be blocked at the stage of light hydrogen and bi-hydrogen (H2/2) would then be formed; the nuclear fusion of deuterium and tritium could not happen and there would be no production of helium (He2/4) and their subsequent deuterotomes. When the neutron losses the electron resulting from a radioactive emission, it once again acquires a proton status even without the electron in the orbital zone. Because this neutron belongs to the structure of a deuterotome, the loss or emission of the electron brings about an increase in the deuterotomic number. The proton then formed receives an electric load, and is leveled downwards due to the higher number of protons in the electronic shell series to which it belongs. As neutron, it has no electric load and does not interfere on the electromotive force (emf) of the series; this occurs only when it loses the electron and acquires a proton status.

The orbital electron of the atom of light hydrogen (H1/1) lies in the K-shell. It is an incomplete atom for the synthesis, because there is no neutron in its structure, as happens to deuterium and tritium. Light hydrogen is inadequate for the synthesis of deuterotomes, unless when it takes on a neutron configuration. However, light hydrogen (H1/1) fails to capture electrons; the capture of electrons by a proton occurs in two different conditions: a) capture of electron by a nuclear proton and b) fusion of two protons thus creating bi-hydrogen (H2/2) and one of these hydrogens captures its own electron, thus forming heavy hydrogen or deuterium (D1/2) in accordance with the following equation:

$$H1/1 + H1/1 \xrightarrow{\text{Fusion}} H2/2$$

$$H2/2 \xrightarrow{\text{Conversion}} D1/2 + \nu + \text{gamma rays } (\gamma)$$

also represented by the equation: $p + e^- \rightarrow n + v$ where **p** is the proton, e^- the electron, **n** the neutron and **v** the neutrino. Light hydrogen is adequate to make up molecules of water (H_2O), acids (HCl, HBr, HNO_3, H_2SO_4) etc., hydrocarbons (CH_4, C_2H_6, C_3H_8 etc.) and hydroxides. Because light hydrogen is incomplete, it is out of the composition of deuterotomes. Should all the atoms of hydrogen be transformed in deuterium (D1/2) or tritium (T1/3), they would be used up during the process of nuclear fusion and there would be no light hydrogen to make up hydrogenated molecules. What little could remain would be formed by deuterium in place of light hydrogen such as D_20, CD_4, DCl, DBr etc., and both the physical and chemical properties would be quite different. Deuterium concentration in water is minimal, that is, 0.7 percent, and is a part of heavy water (D_20).

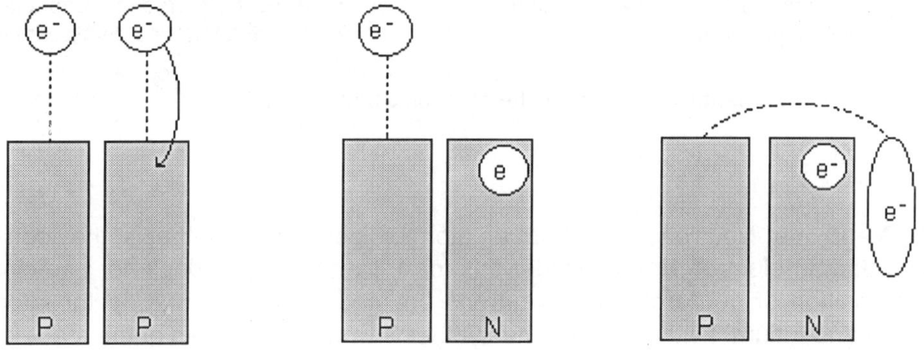

Fig. 5 - Schematic illustration of two atoms of light hydrogen (H1/1) both with their respective electrons in the K-shell making up the transitory molecule of bi-hydrogen (H2/2). In the middle: an electron was captured by its proton, thus forming the neutron. On the right: deuterium with the configuration of an atomic capacitor where the neutron has a dielectric function and both the proton and electron function as plates.

The distance between the electron and the proton is variable. The orbital electrons keep up a distance that is proportional to the square of natural numbers from 1 to 7, that is, 1 for shell K; 4 for L; 9 for M; 16 for N; 25 for O; 36 for P and 49 for Q. The orbital electron with energy on the **s** level, whose proton is juxtaposed, upon being captured by the proton, makes up the neutron, indispensable for the composition of matter. Fig. 5 shows two atoms of light hydrogen with electrons in different energy levels. The electron closer to the proton can be captured and form the neutron. The electron that electrically neutralizes the neutron is juxtaposed to the proton; but not combined with or embodied by it. This electron can roam about or be pulled out that position and then both the proton and the electron can restore their fundamental functions. In their free form, neutrons can at least derive from three different sources:

1) Nuclear fusion of deuterium (D1/2) with tritium (T1/3) produces an unstable helium (He2/5) isotope which disintegrates upon releasing energy, producing beta particles (β), alpha particles (α2/4) and neutron,

2) Nuclear fission of uranium-235 (U92/235) deuterotomes when bombarded by slow neutrons from which two elements of lower deuterotomic numbers derive, namely xenon (Xe54) and strontium (Sr38) in addition to the production of neutrons,

3) Production of radioactive deuterotomes through the bombardment by alpha particles as was the first experimental finding as to the existence of neutrons, called Chadwick's bombardment via high-speed alpha particles over beryl (Be4), as per the following reaction:

Be4/9 + α2/4——————————> C6/12 + n

1.3 – Atoms of hydrogen: composition.

From aggregates of lentons, two routes can be followed: one geared for the production of energy, another for the production of matter. Energy can make up an organized network with interconnected units which in no way form electrons and protons and most likely hard to be detected by current physical means. The production of matter– which is essentially energy–from the atoms of light hydrogen, goes through the production of neutron, that is, the orbital electron closer to the proton approaches it and the potential difference between both then ceases hence producing a structure of unitary mass and null electric load. The appearance of the neutron was vital for matter to be formed. The neutron became a unit capable of clinging onto the proton of light hydrogen atom, causing the latter to become well organized: the atom of heavy hydrogen or deuterium (D1/2). When one other neutron clung onto the neutron of a deuterium, the atom of tritium (T1/3) was formed. The atoms of hydrogen became the basis for deuterotomes to be formed.

Up until now, the origin of matter has not been duly underlined. Atomic Physics itself has no elements to formulate a theory as to its origin. Nearly all theories formulated by Physics thus far are based on fundamental facts as the tangible existence of matter and its physical, chemical, physical-chemical, electromagnetic and radioactive properties. The origin, that is, from were derives the matter which constitutes the well-known universe has not been intellectually probed into at length. The expansion and retraction of the universe, the formation of the stars, the chemical composition of planets and satellites draw a great deal of attention, but not the origin of atoms or matter proper.

This paper brings to light a theory as to the origin of the three basic atoms that make up complex atoms or deuterotomes. The three basic atoms are those of hydrogen in their light, heavy and heavier forms (H1/1, D1/2, T1/3, respectively). These hydrogen atoms are the threshold that demarcates the pre-matter state and matter produced through natural synthesis.

Pre-matter is composed under special form - lentons - in space. This pre-matter originates the atom of hydrogen, the basic component of matter. Matter made up of complex atoms or deuterotomes is the result of synthesis processes from atoms of hydrogen. The following scheme supports the assertion above:

Pre-matter in lentonian space ———>Matter—————> Matter that composes Universe

LENTONS—————>BASIC ATOMS———>COMPLEX ATOMS or DEUTEROTOMES

Based on a theory as to the origin of matter (basic atoms), new insights into the Atomic Theory become a natural consequence. The formulation of a New Atomic Theory displaces concepts and definitions long established or agreed upon. The new word-**deuterotome**-will be defined later on, and the acceptance of deuterium as an atomic capacitor imposes a radical change on the classic atomic theory. Terms and new concepts, such as "electronic furnace", "physical volume-reduction phenomenon", "alpha-appendix", phylogenetic code of "alpha-appendix", genetic pole of deuterotome, chemical pole of deuterotome, synthesis mechanisms, deuterotomic degradation, neutron as dielectric of atomic capacitor, likely pathways for the synthesis of deuterotomes, magnetic tunnel, intradeuterotomic solenoid electrocirculation, potential of neutron-proton contact, capacitive series-equivalent and parallel-equivalent bindings, besides other terms introduced due to an absolute need of clarity. Not all new points resulting from the New Atomic Theory (or Deuterotomic Theory) are highlighted or will be analyzed straight away. The concepts of positive and negative valences match as electric loads of deuterons' valence as high and low voltages (expressed in electron-volts). Because a polarity in deuterotomes can be found from the genetic pole, an opposite pole–the chemical pole–exists as well. Trans-xenon deuterotomes, as more deuterons are added to their structure, take on a round and spiral shape, setting up a pole quite apart from the other. The deuterotome of gold (Au79), for instance, has 25 post-xenon (Xe54) deuterons and a spiral- and round-shaped configuration. The deuterotome of sodium (Na11), conversely, has only one single deuterium and, cannot therefore form a spiral-shaped structure. Below is an outline of a long spiral-shaped deuterotome.

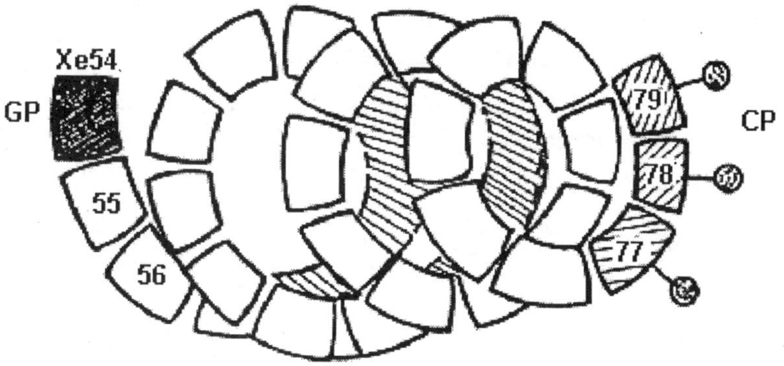

Fig. 6. Simplified scheme showing both the genetic (GP) and chemical poles (CP) of gold deuterotome (Au79) with intermediate atoms of deuterium. The genetic pole is represented by noble gas xenon (Xe54).

1.4–Concepts of Atom and Deuterotome.

One atom of deuterium plus one other of tritium originate helium (He2/4) through nuclear fusion. Thus, this element is composed of two atoms of deuterium, unable to be an atom per se. For that reason, the term deuterotome (G. **deuteros** = second; **tome** = part) started to be used to refer to all elements composed of two or more atoms of deuterium. The term **atom** refers exclusively to the atoms of hydrogen. The word **deuterotome** is more expressive than the term element; for it stands for the composition of the material it refers to. The term element is quite widespread, but it can be likewise used as a synonym just as much as its overall use as an intelligible form. Deuterotomes are defined as organized structures as a result of a synthesis process by "alpha capture" or "neutron capture and emission of beta particle", complying with a phylogenetic code during the synthesis process, in no way, a simple agglomerate of protons and neutrons at random, making up a nucleus.

1.5 – Formation of the Universe and the retraction of the lentonian space.

As lentons cluster or cling together so that electrons and protons can be formed, there occurs a retraction in the limits of the lentonian space moving towards the center where the synthesis of atoms and deuterotomes alike takes place. The formation of stars, planets, asteroids and stellar dusts, all of which make up the constellations, causes lentons to move towards the center of the Universe being formed and transformed, thus creating empty spaces or holes among the Universes then formed. Universes situated nearby can be in touch by way of residual stretches of lentonian space or, otherwise, can be a Universe.

completely apart from all the others and form another one inside a bubble. The following illustration is a simplified view of what has been said above.

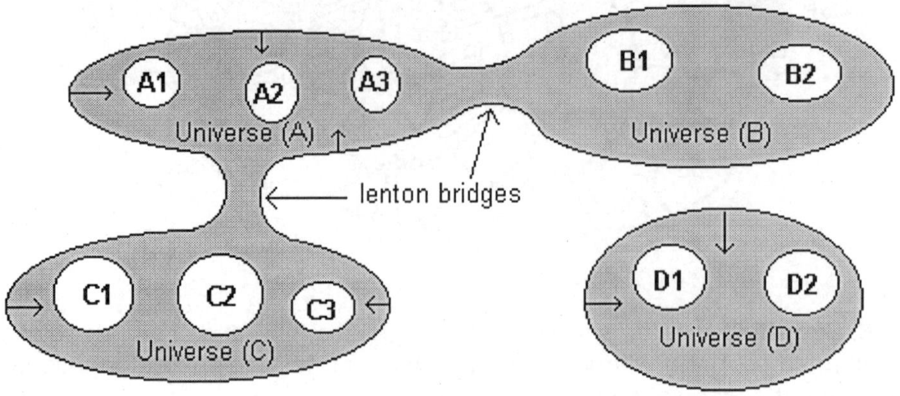

Fig. 7 – Scheme showing interconnected Universes and also set apart from each other inside a bubble.

The phenomena that change the mass of natural lentons into lentons aggregates (Energy and Matter) act endlessly, creating continual Universes in compliance with the same principle. The diameter of each individual Universe is finite and variable. The so-called "worm holes" can correspond to the empty space existing between nearby Universes and "lenton bridges", mass of lentons residues not used in the formation of electrons and protons. Without these connections a Universe could be apart from all the others. The mass that makes up a Universe is finite, albeit individually variable. The volume and mass of stars, planets and stellar dusts will depend on their evolution time and their growth stage. In each stellar unit, the amount of matter that forms their individual mass depends on the initial amount of deuterium and tritium available for the synthesis of deuterotomes by way of nuclear fusion and the consequent production of neutrons, alpha particles and beta particles. When the nuclear fusion process is over as a result of depletion of the atoms of deuterium and tritium, the synthesized matter may have attained such a stage where deuterotomes can consist of few or many atoms of deuterium or even beyond uranium (U92). The phylogenetic lineage of deuterotomes can cease at any stage during the synthesis process, for instance, xenon (Xe54), leaving only the elements formed by deuterotomic numbers to exist up to that element itself-trans-xenon elements excluded. One other example, the syntheses stop at silicium (Si14) stage when elements can be found with only deuterotomic numbers (protons number) up to 14, the others being absent. Likewise, nothing prevents a stellar unit from having had a great number of heavy and heavier hydrogen atoms, maintaining the reactions of nuclear fusion for a length of time beyond the average of other units, thus allowing the synthesis of deuterotomes to overtake that of uranium (U92) or even reach the stage of a noble gas with a deuterotomic number 118. Thus, stellar matters with high radioactivity rates is likely to exist with the emission of alpha and beta particles in addition to radiations by heavy deuterotomes and matter with a low radioactivity rate due to emissions of only beta particles and gamma radiation, resulting from the synthesis process by "neutron capture and emission of beta particles" mechanism and emission of gamma rays.

CHAPTER 2

Atoms of hydrogen as capacitors

2.1 – Industrial capacitors: characteristics

A capacitor consists of two plate-shaped electric conductors separated by an insulating material (dielectric) provided with a capacity called **capacitance**, that is, the capability to store electric load in their electrostatic field. This can be schematically shown as follows:

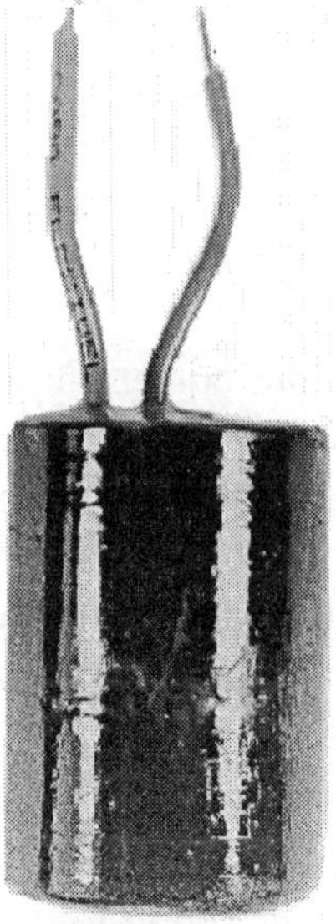

Fig. 8

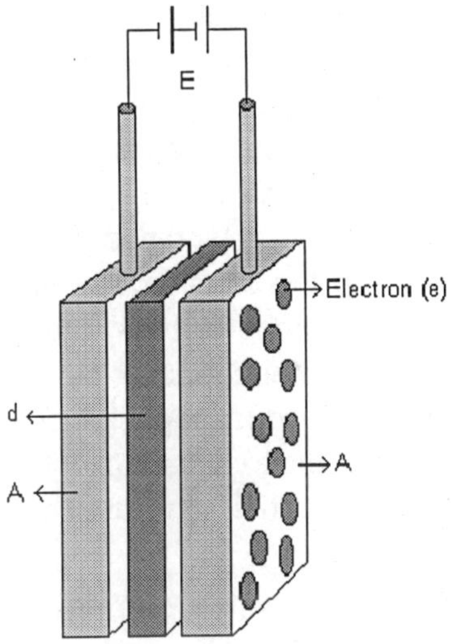

Fig. 9

Fig. 8 – Photo of an industrial capacitor. Each wire corresponds to an internal plate. Insulation (dielectric) exists between the plates to prevent electric current from passing. Example of values: C = ± 10 μF and 250 VAC

Fig. 9 – Simplified scheme of the elements that constitute a capacitor. Two metal plates with area **A**, a dielectric with thickness **d**. The wires connected to the tension source **E** load electrons **e** to one of the plates.

When the two wires A and B are plugged to the source poles (battery) of direct or alternating electric current (industrial electric energy), the electrons are captured by one of the plates and build up to a defined rating called capacity. That feature is known as **capacitance**. The capacitance value is directly proportional to the **dielectric constant** (**k**) of the material that forms the dielectric and also proportional to the plate area (**A**) and inversely proportional to the distance (**d**) between the plates that is occupied by the dielectric. If the area of the plates were expressed in square inches and the distance in inches, then the value 0.2249 would transform the capacitance result into picofarad (pF). The **k** value varies in accordance with the nature of the material that makes up the dielectric: for vacuum this value is equal to 1; for air equal to 1.0006, for paraffined paper 3.5 etc. Thus, the dielectric constant reflects the higher or lower loss of electric load by hysteresis. The higher the **k** value, the higher the capacitance. This also occurs in the area of plates. The shorter the distance **d** between the plates, the higher the capacitance.

A capacitor connected to the tension source gets the supply of a current called **displacement current**. This current develops in the capacitor a counterelectromotive force that is opposed to the tension originating from that source. When the capacitor is totally loaded the cemf will have the same tension as that of the source and the loading current ceases. If the capacitor has been switched off the source, the load will be retained for a long time. The load retention will depend on the amount of **escape current**. Since the capacitor can store electric energy (electrons), a loaded capacitor can function as a tension source and transfer electric discharge to another capacitor or to the resistor of an incandescent lamp, for example. A capacitor can be used separately in an electric system or otherwise be connected to others in order to decrease or increase the total system capacitance. These connections can have series or parallel-bindings. Fig. 10 below shows two elementary capacitors arranged in series-bindings.

In capacitors with series-bindings, the displacement current flows from the loaded plate of one of them to the discharged plate of the other and both share the resulting tension equally. This is Kirchoff's Law: the sum of the tensions of capacitors is equal to that of the source. The total capacitance of series-connected capacitors will always be lower than the lowest individual capacitance. In a circuit with only two capacitors like the one shown in the figure, the total capacitance will be the result of the capacitance of each one divided by the sum of their individual capacitances in accordance with the equation Ct = C1 x C2 / C1 + C2.

The capacitor has a physical property termed **time constant** that is defined as "the time necessary to load or unload it completely should it continue to do so as a reason of initial variation." It can be also defined as "the time necessary to charge the capacitor 63.2% of the maximum tension or discharge 36.8% of that tension when the tension applied to the capacitor is 100V". **RC** is the symbol used for the time constant. In the aforesaid example of 100V, the RC value is 10 microseconds. The RC value, expressed in seconds, is the product of the circuit resistance (R) in ohms and the capacitance (C) in farads in keeping with the equation RC(s) = R (ohms) x C (F). Thus, it can be said that the capacitor recharges more quickly than it discharges. This capacitor property is important for the maximum electric load to be maintained constant at all times.

Fig. 10 – Photo showing two industrial capacitors with individual capacitances of ± 10 μF in series binding. The contact between the two wires levels up the electric charge of both and keeps it unchanged for a long time, but the contact between the loose wires generate a flash and the capacitors are then discharged.

2.2 – Atomic capacitors: characteristics

The synthesis of all matter rests on the atoms of deuterium and tritium. However, before bringing up the synthesis of matter into focus, both the physical and the chemical nature of these atoms should be highlighted.

Deuterium is an atomic capacitor composed of two plates and a dielectric set between them: one is the electron the other the proton; the dielectric is the neutron between them. The proton and the electron consist of lentonians plates and interstitial spaces where free lentons are housed; electrons can be also housed in the interstitial spaces of the protons. The lentons plates that form the structure of the protons can be comparatively regarded as similar to juxtaposed mica plates. The proton is therefore the receiver of free electrons and functions as the plate that stores electric load. Electrons or lentons entering the protons interstice represent a rise in the electric load. The capacitance of the atomic capacitor can be expressed in electron-farad (eF). Fig.11 represents the structure of an atomic capacitor composed of proton **(P)**, electron **(e)** and neutron **(n),** between them. In Fig. 12 the proton plate is provided with units that represent electrons in the interstitial spaces of lentonian plates.

Fig. 11

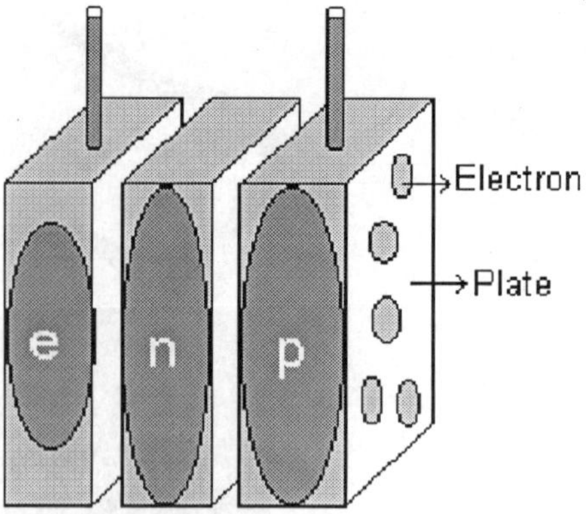

Fig. 12

Fig. 11 – Photo of a simplified illustration of the atomic capacitor structure (deuterium) composed of proton plates (P) and electron (e), and the dielectric neutron (n) by intermediation.

Fig. 12 – Simplified scheme of the atomic capacitor (deuterium), composed of plates **P** and **e**, and neutron **n** as dielectric. The plate of proton P captures electrons.

The neutron, which works as a dielectric or insulation for the plates, should be equal to 1 (k = 1). The distance separating plates **P** and **e** corresponds to the distance from the electron to the proton, intermediated by the dielectric; in the case of deuterium it is made up of one neutron and in the case of the tritium of two neutrons; the value of **d** is equal to 1 and 2, respectively.

In order for the deuterium atom to be able to function as an atomic capacitor, it is also necessary that the proton function as a plate and the orbital electron as the other. Between the two plates there must be a dielectric; that is, an electric insulation. Thus, the neutron should be the dielectric for the atomic capacitor. The absence of neutron in the atom of light hydrogen (H1/1) does not enable it to feature an atomic capacitor for the synthesis of matter. The fact that the deuterium atom possesses one neutron and tritium two neutrons enables matter to have units with different capacitances. Because light hydrogen cannot be a capacitor and because deuterium and tritium alike have different capacitances, these atoms are extremely important for the constitution of the elements and molecules. Alternatively, the interplate space could be empty and the dielectric in that case would be a vacuum. Air is made up of matter and in no way can it be the dielectric for an atomic capacitor. However, the dielectric constant of the neutron should be the same as that of vacuum, that is, equal to 1 (k = 1). The A surface of the plate should have the value of the hemisphere with a diameter of 1. The distance between the plates should correspond to the thickness of a neutron and should be measured as 1; d would take the value of 1 for deuterium and 2 for tritium. In the capacitance formula when the values are expressed in inches, this value can be multiplied by the constant 0.2249 and the result expressed in picofarad (pF). The capacitance is calculated as per the equation **C = 0.2249 x k x A/d**, where **C** is the capacitance (in Farad); k is the dielectric constant of material (neutron in the example); **A** is the plate surface (expressed in square inches); **d** is the distance between the plates (expressed in inches); 0.2249 is a constant that transforms Farad/inch into pF/cm. As for the atomic capacitor, the result will be expressed in electron-farad. Calculations will be shown in due course. The photo shows three atoms of hydrogen with their characteristics.

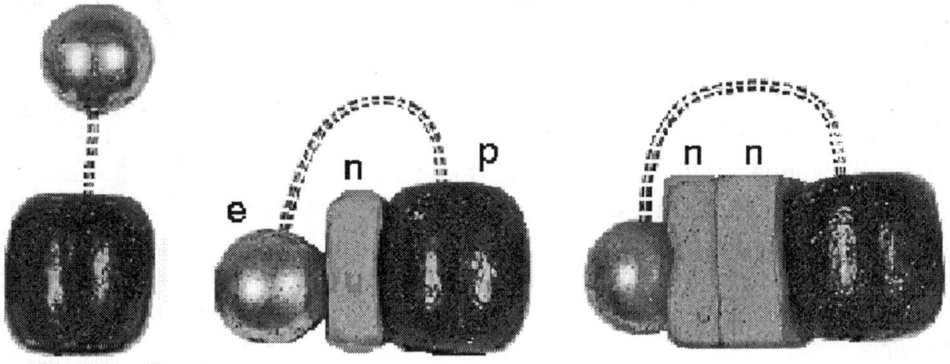

Fig. 13. Photo illustration of the three hydrogen atoms, where **P** is the proton, **e** the electron and **n** the neutron. 1) – Light hydrogen with its orbital electron in the K-shell, and distance 1 between the proton and the electron; there is no neutron. 2) – Heavy hydrogen (deuterium) with one neutron between the proton and the electron. 3) – Heavier hydrogen (tritium) with two neutrons between the proton and the electron.

According to the capacitance equation if the distance between the plates increases, the capacitance then decreases. Accordingly, because deuterium has one neutron and tritium two, the capacitance of the latter is lower. Capacitors with low capacitance can withstand very high tensions. Comparatively, it can be said that the capacitor that covers the strip between 1 pF and 0.01 mF can be used under high electric tension, as high as 30,000 V.

The following common and differentiate characteristics between the elementary capacitor (industrial) and the atomic capacitor should be highlighted: 1) the types of elementary capacitor can be: **fixed** and **variable**; the tension source can be of **direct current** and **alternate current**; the number of plates can be **double** or **multiple**; the dielectrics can be of air, vacuum, mica, ceramic, paraffined paper etc.; the bindings can be in **series** and **parallel**; 2) the atomic capacitor can be also of a fixed type-when it consists of an atom, deuterium or tritium-and a variable type, when it is composed of deuterotomes that undergo the physical phenomenon of volume-reduction, as is the case of noble gases; the tension source is the receiver of free electrons, free lentons and quanta; the number of plates can be double (deuterium) and multiple (noble gases and deuterotomes in general); the dielectric is the neutron and it may be single (in deuterium) or double (in tritium); the arrangement can be series-equivalent between the pair of deuterons that make up the structure of deuterotomes ranging from the noble gas to another in the K, L, M, N and O-shells, that is, from the deuterotome of helium (He2/4) up to that of xenon (Xe54). The series-equivalent arrangement brings about the physical phenomenon of volume-reduction upon reaching the deuterium that completes the deuterotome of a noble gas. The parallel-equivalent arrangement among the deuterons that range from xenon (Xe54) to radon (Rn86), thus linking one deuterium to another, causes a concentration of electrons at a point close to the so-called genetic pole because the volume-reduction phenomenon fails to occur with the synthesis of radon (Rn86), thus forming the "electronic furnace". Only the 32 electrons ranging from deuterium D55 to D86 participate in the formation of the "electronic furnace"; the 6 other electrons belong to the uranium deuterotome (U92).

Thus, all the physical features of the industrial capacitor, which could be analyzed in keeping with its macrostructure, are contained in the atomic capacitor both as a capacitive unit and types of tension sources, number of plates, dielectrics and arrangements. One of the most extraordinary facts found in atomic capacitors is that they can have a wide range of capacitances and may have their values increased when noble gases undergo a dramatic reduction through the physical phenomenon of volume-reduction in distance **d** between the plates. No less significant is the capability to make such arrangements in series-equivalent and parallel equivalent, thus determining the volume-reduction phenomenon in the first arrangement and the formation of the "electronic furnace" in the second one.

Therefore, the number of plates in each noble gas is similar to the number of protons that compose them: helium (He2/4) has 2 plates; neon (Ne10) has 10; argon (Ar18) 18; krypton (Kr36) 36; xenon (Xe54) 54; and radon (Rn86) 86.

In the physical phenomenon of volume-reduction, the noble gases neon and argon have their volumes reduced twice the volume of hydrogen while the other noble gases have their volumes reduced to the volume of the atom of hydrogen. This fact also causes the capacitance of the deuterotome to vary because distance **d** between the plates is greater in noble gases neon and argon than would be expected, should the reduction rate rise. This fact is also important because it is applied to the capacitors (noble gases) that form the deuterotome's genetic pole. The deuterotome that composes the genetic pole must have a high capacitance, which is attained through the reduction of distance **d**. The higher the number of plates and the lower the **d** value, the higher the capacitance value.

The deuterotome's synthesis can be dependent on the capacitance of noble gas, possibly one of the basic requirements that can cause an alpha particle to be captured by that gas. The electric load of the first deuterium that is found linked to the noble gas is as high as the greater the number of plates of the noble gas that forms the deuterotome's genetic pole. Thus, the electric load of the deuterotome of cesium (Cs55) is higher than that of the deuterotome of sodium (Na11), because xenon (Xe54) has 18 plates and neon (Ne10) only 8.

In a series-equivalent capacitive circuit, the displacement current flows through each part of the circuit and the tension applied is divided individually among the capacitors. Thus, the capacitance decreases as alpha particles are added to the previous deuterotome. The capture of alpha particles corresponds to the increment of two series-connected capacitors. This increment occurs upon every new capture of an alpha particle. The equation for the capture of alpha particles can be shown as follows:

$$Ne10/20 \xrightarrow{+\alpha 2/4} Mg12/24$$

The electric load of neon is 8 eV, which will be equally shared by the two deuterium atoms that come around after the neon in the magnesium deuterotome. Accordingly, each deuteron D11 and D12 has 4 eV in accordance with Kirchoff's Law that states that the sum of tensions in capacitors is the same as that of the source tension. As per the equation Et = E1 + E2 where Et is the source tension which is the genetic pole in the example given, composed of neon equal to 8 eV; by using the equation to calculate both the value of E1 and E2 it will be equal to 4 eV.

2.3–Neutrons as dielectric plates of atomic capacitors

It can be inferred that the neutron is a dielectric whose electrons escape is almost null. In other words: the neutron can be considered the insulating plate of dielectric hysteresis or dielectric loss almost null. As for dielectrics, vacuum is the one that has lower dielectric loss through hysteresis. Such being the case, should the atomic capacitor have a loss almost null, the **power factor** of that capacitor would be equal to zero and the capacitor would resume its total load.

The single atomic capacitor is a fixed capacitor at a given moment, that is, it possesses a fixed capacitance value that cannot be reset. What varies in the making of matter is the number of dielectrics (or the thickness between plates) in deuterium and tritium. For that reason, during the process of deuterotomes synthesis, the existence of deuterium (D1/2) and neutrons to make up the phylogenetic lineage is important in order to carry on the synthesis or interrupt its progression. As a result, some lineages give the synthesis a rest (stop), still with few deuterium atoms in their structure.

In short, an atomic capacitor is made up of two main plates, one of proton and one of electron, respectively. For its turn, a proton plate is composed of x.1,836 lenton plates while an electron plate consists of x.1 lenton plates; while a fixed type capacitor, it has either a single or a double dielectric between the plates and can function with other series-equivalent or parallel equivalent atomic capacitors. The photo that follows illustrates both deuterium and tritium as capacitors, the former with one dielectric (neutron), and the latter with two.

2.4 – The importance of the physical phenomenon of volume-reduction of noble gases in the capacitance values.

Because an atomic capacitor can be of a fixed type at a certain time or condition, this fact does not prevent it from increasing its capacitance. When two deuterium atoms are together in synthesis to form the deuterotome of helium (He2/4), the volume is reduced by half, that is, the helium volume thus formed then acquires the volume of a proton. According to Avogadro, regardless of its mass number, every gas occupies the same volume with the same number of deuterotomes. Thus, two deuterotomes-gram of helium (He2/4) fill up the same volume of 54 deuterotomes-gram of xenon (Xe54).

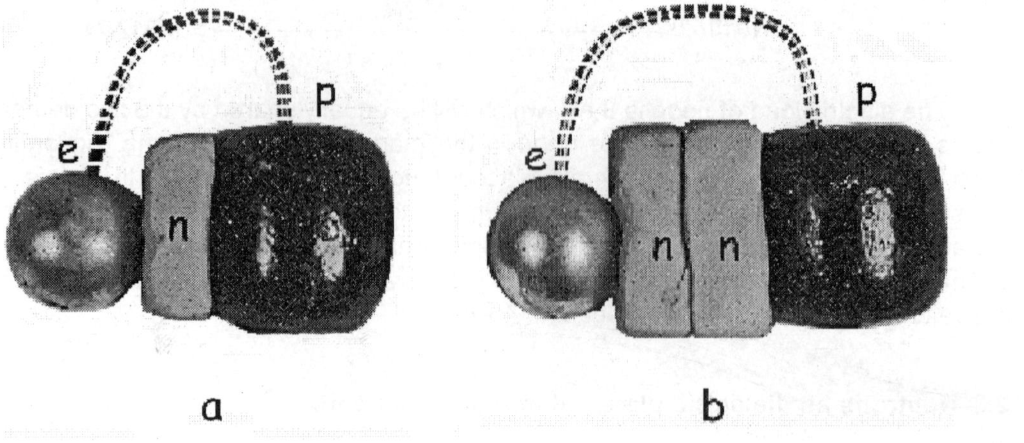

Fig. 14. Photo illustration of deuterium **a** and tritium **b** highlighting the position of the neutrons (dielectrics) between proton **(P),** and electron **(e),** as well as the distance **(d)** between the plates.

The volumetric-reduction of helium deuterotome, as well as that of all other noble gases is of great physical importance, for it reduces distance **d** between the plates and brings about a significant rise in the capacitance value. Thus, because the atomic capacitor is not of the rotor-stator type (variable capacitor), it can be considered the perfect capacitor, since it is a fixed type capacitor under a certain condition and variable type under another. The capacitance value of noble gases increases for each new deuterotomes series (He2, Ne10, Ar18, Kr36, Xe54 and Rn86). Thus, the atomic capacitor, from the physical point of view, is the one that meets the needs determined by the deuterotomic phylogenetics.

In a deuterons sequence–part of the deuterotome structure–they can be series-equivalent or parallel equivalent connected, wholly or in part. In series-equivalent bindings, deuterons are connected when they are made up of deuterotomes between the first deuterium after the noble gas and the last deuterium before the next noble gas. Consequently, these bindings occur in the interval ranging from lithium (Li3) to fluorine (F9), from sodium (Na11) to chlorine (Cl17), from potassium (K19) to bromine (Br35) and from rubidium (Rb37) to iodine (I53). In trans-xenon deuterotomes (Xe54), the bindings between deuterons are parallel equivalent. When the series-equivalent connections reach the deuterium that completes the deuterotome of noble gas (helium, neon, argon, krypton and xenon) all of the series undergo a volumetric reduction, producing a significant increase in the capacitance and raising their density.

As regards the basic functions that an elementary capacitor (industrial) can have, depending on its constitution and the place where it is used are those listed below:

1) Electric current blocked. Should the circuit have both alternating (AC) and direct currents (DC) and should only alternating current be needed, a block-in capacitor prevents the outflow of direct current.

2) Prevention of sparking formation. The electric flash that appears on the relay contacts or the contact keys can be prevented or minimized by a capacitor installed through the relay contacts or some other point subject to sparking resulting from the circuit opened or closed. The valences that are opened or closed up in even numbered deuterons, self-neutralized, may call for the sparking-formation suppressor effect by the capacitor, as, for example, chlorine deuterotome (Cl17) that can have one, three, five or seven valences, depending on how it is stimulated,

3) Filtration of the excess of electric tension that leaves the source and goes to the electric load arrester. If the receiver is prepared to work with 100 V tension and the source tension varies upwards, for example 150 V, then the capacitor can filter out that excess tension, thus avoiding damaging the receiver. During that filtration, the capacitor produces insulation for electric tensions that should not be exceeded. The filter capacitor maintains the current tension without oscillations and eliminates the fluctuating tension by the capacitor's capability in fighting back any tension variation.

The electric load of the atomic capacitor is stored in the proton plated-structure. Therefore, in the pathways for the synthesis of deuterotomes, when an alpha particle (α) is captured, the deuterotome then formed gets two other capacitors because the alpha particle has two deuterium atoms in its structure. When the deuterotome captures a neutron, this neutron emits a beta particle (β), acquires a proton states and becomes the increment of a capacitor in the structure of the new deuterotome with the deuterotomic number being added by 1.

Thus, the increment of deuterons to deuterotomes is regarded as an increment to the number of capacitors. Deuterium, considered individually, is a capacitor of fixed capacitance but the addition of one more neutron causes its capacitance to drop to a lower value owing to the increase experienced by variable **d**. When capacitors (deuterons) are series-equivalent connected and fail to undergo the physical volume-reduction phenomenon, both their individual and total capacitances are reduced. When capacitors (deuterons) are parallel equivalent connected only their total capacitance is increased. They do not undergo any physical phenomenon of volume-reduction and give rise to the formation of electrons agglomerate that makes up the "electronic furnace" in the uranium deuterotome (U92).

As a result of the physical phenomenon of volume-reduction, noble gases acquire a super-density, as is the case of xenon (Xe54) with 131.3 g/cm^3 and krypton (Kr36) with 83.7 g/cm^3, for example. Each noble gas has a different capacitance due to the variation in the number of plates of each deuterotome and the physical phenomenon of volume-reduction that differs for neon (Ne10) and argon (Ar18) in comparison with other noble gases. The area of the plates does not vary; the capacitance variation is dependent on the reduction rate of distance **d**.

Whereas deuterium (D1/2) is a typical capacitor provided with two plates and one dielectric, respectively represented by the proton/electron and the neutron, it should have the unit value. The only fixed value in that case is the dielectric constant (k) of the neutron, which holds good for all conditions, since it is the only dielectric possible. Because the proton's electric load is the same as that of electron, the atom capacitance can be expressed in electron-farad (eF). Since the proton mass (pm = 938.28 MeV/c^2) is almost similar to that of neutron (nm = 939.57 MeV/c^2), the diameter of both also tends to have values close to each other. For the purpose of calculation, the diameter of both can be adopted as similar to that of 1 proton (1 p). Once the parameters for the calculation of the unit capacitance have been established, there remains only the calculation of the proton plate area. Because proton is a structure with reasonably spherical shape, the area of the hemisphere becomes the plate area for the single capacitor. Thus, in the equation below the area of hemisphere **(A)** can be calculated by regarding the radius of the sphere as equal to 0.5 p (half diameter of the proton). The basic capacitance equation **(C)** is as follows:

$$C = \frac{k \cdot A}{d} \quad \text{and} \quad A = \frac{4\pi \cdot r^2}{2}$$

By substituting r:

$$A = \frac{4\pi \cdot (0.5p)^2}{2} \therefore \quad A = \frac{4\pi \cdot 0.25p^2}{2} \therefore \quad A = 1.5708p^2$$

Thus, the value of A is a constant that corresponds to the area of the proton hemisphere. In the basic capacitance equation of atoms and deuterotomes, the A value will always be the same. However, the value of A should be multiplied by the number of plates (or protons), the value N–1 not being adopted in the case of a parallel-equivalent binding because the plates do not interweave and only remains linked to each other. The value of **d** should also be multiplied for the number of neutrons (dielectrics) because it corresponds to 1 **p**. The equation that enables an atom capacitance or a deutrerotome to be calculated is as follows:

$$C = \frac{k \cdot A \cdot np}{d \cdot nn}$$

And by substituting A for its value

$$C = \frac{k \cdot 1.5708p^2 \cdot np}{d \cdot nn}, \text{ where} \tag{1-1}$$

C = Capacitance of the unit, expressed in electron-farad (eF),

k = Dielectric constant of the neutron, expressed in 1 eF.p^{-1},

A = Area of the hemisphere of the proton, expressed by 1.5708 p^2, which is the constant value,

np = Number of protons that compose the capacitor (or number of plates),

d = Diameter of the neutron (dielectric), equal to 1 p,

nn = Neutrons number (dielectrics) that compose the capacitor (or that are a part of the deuterotome's composition).

To calculate the deuterium capacitance by means of the equation, the following values are found: k = 1 eF.p^{-1}, A = 1.5708 p2; np = 1 p; d = 1 p; nn = 1 p. If we substitute the values in the equation, we have:

$$C = \frac{1eF \cdot p^{-1} \cdot 1.5708p^2 \cdot 1p}{1p \cdot 1p}$$

where

C = 1.5708eF

The capacitance of the heavier hydrogen atom, namely tritium (T1/3), has a lower value where an additional dielectric exists since the distance between the plates has doubled. By adopting the said equation where **nn** is equal to 2 the capacitance value equals 0.7854 eF, which is also the smallest capacitance value between the atoms and deuterotomes, excluding the atom of light hydrogen (H1/1) and the neutron that are not considered capacitors. Among heavy deuterotomes, the one having the smallest C value is radioactive lead-214 (Pb82/214), of radium family (Ra88), averaging 0.9758eF. Uranium-238 (U92/238) has a capacitance of 0.9898eF.

The highest capacitance values are encountered in noble gases as a result of the physical volume-reduction phenomenon they go through. By applying the equation (1-1) and because the volume-reduction of neon (Ne10) and argon (Ar18) fail to occur until the volume of an atom of hydrogen is reached like all the others, but rather until twice this volume, the values are then as follows:

CHe2/4	=	1.570	eF
CNe10	=	7.854	eF
CAr18	=	28.270	eF
CKr36	=	56.540	eF
CXe54	=	84.820	eF
CRn86	=	135.080	eF

The capacitance of a certain noble gas is identical to all of its isotopes because the physical volume-reduction phenomenon nullifies any variaton in **nn** values, making it equal to 1p or 2p depending on the noble gas considered for analysis.

The capacitance of the isotopes of an element, noble gases excluded, is lower in the isotope having the highest mass number. In the table below, the isotopes of deuterotomes of zirconium (Zr40), polonium (Po84) and astatine (At85) are illustrated as examples of variable capacitance values for the isotopes; also indicated is the percentage or their individual half-life.

Thus, the lower capacitance for an isotope with higher mass number is the consequence of the increase in the number of neutrons (dielectrics) with an increase in the product of the divider in the capacitance equation. Both in the synthesis mechanism where neutrons can be received and in that of degradation due to neutron emission, the **nn** values undergo higher or lower variations, respectively, causing the capacitance values to drop or rise. The transformations in the schemes (the first two) show neutron capture with declining capacitance and neutron emission with increasing capacitance:

Se14/80 ---+n---> Se34/81 ---- -β -γ ----> Br35/81

Pb82/208 ---+n---> Pb82/209 ---- -β -γ ----> Bi83/209

Ba56/141 ---−n---> Ba56/140 ---- -β -γ ----> La57/140 ---- -β -γ ----> Ce58/140

Isotopes	% or average life	Capacitance (in eF)
Zr40/88	83.4 days	1.3090
Zr40/90	51.5%	1.2566
Zr40/91	11.2%	1.2320
Zr40/92	17.1%	1.2083
Zr40/93	1.5×10^6 years	1.1855
Zr40/94	17.4%	1.1635
Zr40/95	64 days	1.1424
Zr40/96	2.80%	1.1220
Po84/207	5.7 hours	1.0727
Po84/208	2.9 years	1.0640
Po84/209	102 years	1.0555
Po84/210	138.4 days	1.0472
At85/209	5.4 hours	1.0767
At85/210	8.3 hours	1.0681
At85/211	7.2 hours	1.0596

In the table below the calculated values of capacitance are shown for each element formed in transformation sequence, as well as the values of total binding energy (Tbe) for each deuterotome and the binding energy per nucleon (Ben). Nucleon is a term that designates neutron and proton independently.

Elements	Total binding energy	Binding energy per nucleon	Capacitance (in eF)
Se34/80	696.872	8.710	1.1610
Se34/81	703.574	8.686	1.1363
Br35/81	704.377	8.696	1.1951
Pb82/208	1,636.549	7.867	1.0222
Pb82/209	1,640.390	7.848	1.0142
Bi83/209	1,760.620	8.425	1.0347
Ba56/141	1,177.266	8.349	1.0348
Ba56/140	1,170.510	8.360	1.0472
La57/140	1,173.485	8.382	1.0787

It can also be observed that the emission of a beta particle (β-)–by decreasing the **nn** value and increasing the np value (in the capacitance equation)–causes the capacitance to rise to a value higher than that in the deuterotome that gave rise to the mechanism of synthesis type “neutron capture and emission of beta particle (β)”.

The bending energy per nucleon, expressed in mega-electron-volts (MeV), keeps up the same type of value variation–upwards or downwards–that occurs with the capacitance. However, the same thing does not happen to the total binding energy of the deuterotome. The isotopes with higher mass number have smaller capacitance values as well as lower binding energy per nucleon and higher values of total binding energy. The table below illustrates the comparative values between uranium isotopes:

Isotopes	Total Binding Energy (in MeV)	Binding energy per nucleon (in MeV)	Capacitance (in eF)
U92/235	1,783.88	7.591	1.0105
U92/238	1,802.00	7.571	0.9898

Clearly enough, a neutron entering the structure of a deuterotome has its capacitance reduced, but those that maintain the ratio 1 proton: 1 neutron will always have the maximum capacitance of 1.5708 eF, as is the case of lithium-6 (Li3/6), carbon-12 (C6/12), nitrogen-14 (N7/14), oxigen-16 (O8/16), magnesium-24 (Mg12/24), silicium-28 (Si14/28), sulfur-32 (S16/32) and calcium-40 (Ca20/40). Isotopes of these elements with higher number of masses have lower capacitances, e.g., lithium-7 (Li3/7), whose value is 1.1781eF. The demonstrative picture that follows shows the capacitance values for some elements to highlight the decrease of those values up to uranium-238. However, a few elements possess lower values than that of uranium-238, as is the case of Pb82/214 with 0.9768eF and Th90/234 with 0.98175eF. In trans-uranium elements, the capacitance values are increased.

The capture of alpha particles ((2/4) in the synthesis mechanism of deuterotomes beginning from helium-4 (He2/4), produces deuterotomes with similar capacitance values, as in the phylogenetic lineage that follows:

$$He2/4 \xrightarrow{+\alpha} B4/8 \xrightarrow{+\alpha} C6/12 \xrightarrow{+\alpha} O8/16 \xrightarrow{+\alpha} Ne10/20 \xrightarrow{+\alpha} Mg12/24$$

Any synthesis sequence set about in conformity with the same mechanism, but whose stem deuterotome is exceeded by one or more neutrons, thus breaking up the ratio 1 proton: 1 neutron, will originate deuterotomes with increasing capacitance values. In the phylogenetic lineage ranging from ruthenium-100 (Ru44/100) to tin-112 (Sn50/112), for example, the capacitance values are increasing. The sequence scheme for capturing of alpha particles ((2/4) is shown below:

$$Ru44/100 \xrightarrow{+\alpha} Pd46/104 \xrightarrow{+\alpha} Cd48/108 \xrightarrow{+\alpha} Sn50/112$$

The capacitance values of each element, expressed in eF, are as follows: 1.2342, 1.2458, 1.2566 and 1. 2667:

Elements	Capacitance (in eF)	Elements	Capacitance (in eF)
D1/2	1.5708	Cr24/52	1.3464
T1/3	0.7853	Mn25/55	1.3090
He2/3	3.1416	Fe26/58	1.2760
He2/4	3.1416	Co27/59	1.3253
Li3/6	1.5708	Ni28/58	1.4660
Li3/7	1.1781	Br35/79	1.2495
Be4/9	1.2566	Kr36/84	56.5480
B5/11	1.3090	Rb37/85	1.2108
C6/12	1.5708	I53/127	1.1250
N7/14	1.5708	Xe54/132	84.8232
O8/16	1.5708	Cs55/133	1.1076
F9/19	1.4137	Hg80/202	1.0300
Ne10/20	7.8540	Rn86/222	135.0880
Na11/23	1.4399	Fr87/223	1.0048
Mg12/24	1.5708	Ra88/226	1.0016
Al13/27	1.4586	Ac89/227	1.0130
Si14/28	1.5708	Th90/232	0.9955
P15/31	1.4726	Th90/234	0.9817
S16/32	1.5708	Pa91/233	1.0066
Cl17/35	1.4835	U92/235	1.0105
Ar18/36	14.1372	U92/238	0.9898
K19/40	1.4212	Np93/239	1.0005
Ca20/40	1.5708	Cf98/251	1.0061
V23/51	1.2903	Rf104/261	1.0405

Binding energy – A fact of significant importance in the field of Nuclear Physics is the binding forces among nucleons, that is, the energy that links a neutron to a proton or a neutron to another neutron, considered individually. Calculations have demonstrated that this energy is lower among the nucleons of heavy deuterotomes. In order to obtain the value of the binding energy per nucleon (ben), first of all it is necessary to calculate the total binding energy of the deuterotome (tbe), which can be found by the following equation:

$$\mathbf{Tbe = (N \times mn + Z \times mp - A) \times c^2}$$

where **N** is the number of neutrons that take part in the deuterotome structure, **mn** the mass of the neutron expressed in **u** (unit of atomic mass), **Z** the number of protons that take part in the deuterotome structure, **mp** the mass of the proton, A the mass of the deuterotome (**N + Z**) and c^2 the square speed of light. Accordingly, in keeping with the equation of the Einstein's special theory of relativity,

$$E = m \times c^2, \text{ or } c^2 = E/m$$

where **m** is equal to 1u (unit of atomic mass) and **E** (kinetic energy) equal to 931.5 MeV. By substituting that value in the c^2 equation, we have: c^2 = 931.5 MeV/u. Likewise, by substituting the C^2 value in the Tbe equation, we have:

Tbe = (N x mn + Z x mp – A) x 931.5 MeV/u

In the process of nuclear fission of uranium-235 (U92/235) with slow neutrons acting as atomic projectiles, the deuterotomes of **heavy fraction** and **light fraction** resulting therefrom can have their values of total binding energy and binding energy per nucleon calculated as per that equation. The deuterotomes of uranium-235 bombarded with neutrons undergo nuclear fission in a chain reaction, splitting the proton-neutron binding between D55 and D56 deuterons, resulting in the deuterotomes of cesium (Cs55) and rubidium (Rb37) or between D56 and D57 deuterotomes, resulting in barium (Ba56) and krypton (Kr36) deuterotomes - in both cases - heavy fractions and light fractions, respectively. The following equation shows the former reaction referred to above.

U92/235 ————————→ Cs55/141 + Rb37/93 + 2n

By applying the respective values to the Tbe equation then we have: uranium-235: N = 143 neutrons; mn = 1.008665u; Z = 92 protons; mp = 1.007825u; A = 235.043925u (mass number of that isotope). By substituting the values in the equation:

Tbe = (143 x 1.008665u + 92 x 1.007825u – 235.043925u) x 931.5 MeV/u

Tbe = 1,783.88 MeV

The binding energy per nucleon (ben) corresponds to the result obtained from the division of the total binding energy (tbe) by the sum of neutron (N) and the number of the protons (Z), globally designated as nucleons. Thus, for uranium-235 deuterotome we have:

$$\text{Ben} = \frac{\text{Tbe}}{\text{N}+\text{Z}} \therefore \quad \text{Ben} = \frac{1.783.88\ \text{MeV}}{143+92} \therefore \quad \text{Ben} = 7.591\ \text{MeV per nucleon}$$

for cesium-141 (Cs55/141), the Tbe calculation is as follows:

Tbe = (86 x 1,008665u + 55 x 1,007825u – 140,908279u) x 931,5 MeV/u

Tbe = 1,180.476 MeV

The calculation of Ben for Cs55/141 (1,180.476 MeV ÷ 141) accounts for 8.372 MeV per nucleon.

for rubidium-93 (Rb37/93), the Tbe calculation is as follows:

Tbe = (56 x 1.008665u + 37 x 1.007825u – 92.906477u) x 931.5 MeV/u

Tbe = 808.8 MeV

and for Ben (808.8 MeV ÷ 93) the value is equal to 8.696 MeV per nucleon.

In the reaction of nuclear fission of uranium-235, with the formation of heavy-fraction (Ba56/140) and light-fraction (Kr36/93) as follows:

U92/235 ---------------+n---------------> Ba56/140 + Kr36/93 + 2n

As a result of the calculations made in accordance with the equations shown above, the table for the values encountered is as follows:

Elements	Total binding energy (in MeV)	Binding energy per nucleon (in MeV)	Capacitance (in eF)
U92/238	1,802.00	7.571	0.9898
U92/235	1,783.88	7.591	1.0105
Ba56/140	1,170.51	8.360	1.0472
Kr36/93	809.50	8.705	0.9928
Cs55/141	1,180.40	8.370	1.0045
Rb37/93	808.8	8.696	1.0378

Some examples show the Tbe and Ben values per deuterotomes and nucleon, respectively.

Elements	Tbe	Ben	Elements	Tbe	Ben
D1/2	2.224	1.112	Co27/58	517.150	8.765
T1/3	8.482	2.827	Ga31/69	602.016	8.724
He2/4	28.290	7.072	Kr36/84	732.265	8.717
Li3/7	39.244	5.606	Sn50/120	1,020.560	8.504
Be4/9	58.164	6.4.62	I53/127	1,072.580	8.445
B5/11	76.206	6.927	Xe54/132	1,112.740	8.427
C6/12	92.162	7.680	Cs55/133	1,118.550	8.410
N7/14	104.659	7.475	La57/139	1,164.560	8.378
08/16	127.620	7.976	Hg80/202	1,595.190	7.896
Ne10/20	160.647	8.032	Bi83/209	1,640.250	7.848
Mg12/24	198.255	8.260	U92/235	1,783.880	7.591
Ar18/40	343.813	8.595	U92/238	1,802.000	7.571
Fe26/58	492.300	8.791			

The increment of a neutron in the deuterotome structure represents an increase in the total binding energy and a decrease in the binding energy per nucleon, thus showing that the binding energy between one proton and one neutron or between two neutrons is weaker. This is likely to be one of the factors that contribute to the radioactive degradation of heavy deuterotomes ranging from uranium (U92) to bismuth (Bi83).

2.5–Atomic capacitor: matter's vital unit

The atom is an electric unit. As such, it must have a form and a function to enable its energetic load to vary. The model of Rutherford-Bohr and Thomson for the atom of light hydrogen (H1/1) can be the basis for better understanding. Nevertheless, to adopt the same atomic model for all the other atoms from light hydrogen (H1/1) to uranium-238 (U92/238) and all the trans-uranium, transforms the nucleus of these atoms as simple clusters of protons and neutrons distributed at random, despite Meyer theory that sets protons and neutrons with shell distribution (magic numbers).

The new concept, in which the proton can be composed of lenton plates and capable of storing energy inside, has conferred upon that unit the feature of a capacitor plate. Thus, the atom of heavy hydrogen, namely deuterium (D1/2), which consists of a proton and a neutron, clearly shows that neutron can work as a dielectric, that is, an electric insulating due mainly to the fact that it has no electric load. Compared to a plate (proton) capable of storing energy and a dielectric capable of isolating energy from flowing through, the electron is now seen as the likely second plate of a capacitor, the **atomic capacitor**. It became thus evident that the plate (proton) with a mass 1.836 times greater than the other plate (electron) would be the receptor plate for the electric load and its storage. Accordingly, deuterium is the single capacitor of matter or the single atomic capacitor. Based on the knowledge that deuterium is an atomic capacitor, a new atomic theory could be established by taking into account the physical properties of capacitors as regards the number of plates, dielectric thickness, dielectric constant, series-equivalent and parallel-equivalent bindings, fixed types and variable types in relation to the thickness of the dielectrics. This new theory being unfolded certainly leads towards unrivalled findings. The first steps have been taken to cast light on the existence of the "electronic furnace", the physical volume-reduction phenomenon, the potential bindings of capacitors in series-equivalent and parallel-equivalent connections, the concept of a genetic pole and a chemical pole for deuterotomes, the increase and decrease of capacitance values, the notion of phylogenetic lineage of deuterotomes, the likely pathways for the synthesis of deuterotomes through its basic mechanisms, besides other less important notions.

An industrial capacitor, composed of its basic elements, with two plates and a dielectric among them, has the following four basic functions:

1) It stores electric load in the receptor plate,

2) If there is an electric load of direct and alternating current, it can block one of them and let the other flow through,

3) Capable of storing electric load, it can reduce spark formation or even prevent it from breaking out in the points of contact of manual keys or automatic relays, avoiding fire formation or flashes, reducing or preventing the wear and tear of the part where the contact is processed,

4) It filters off the excess of electric tension (voltage) which flows onto to the next electric unit (transistor, photoelectric cell, resistor of any nature etc.), not allowing tension

to pass beyond the rate that the circuit can bear (e.g.: in an incoming load of 150V, where the unit can only withstand 110V, the capacitor allows this voltage to pass through, but retains the remaining 40V).

An atomic capacitor or its combination in series-equivalent or parallel equivalent bindings, equipped with its mechanisms to increase or decrease the capacitance values, can perform the following tasks either individually or by forming a structural system:

1) - Store electric load in the receptor plates made up of protons,

2) - Divide or distribute electric load onto the deuterons that make a part of the deuterotomes structure synthesized by alpha particles or neutron capture,

3) - Store high electric load in deuterotomes of short structure like sodium (Na11), potassium (K19), rubidium (Rb37) etc. and low electric load in deuterotomes of long structure such as fluorine (F9), chlorine (Cl17), bromine (Br35) etc.,

4) - Avoid or reduce the likelyhood of spark formations among the valences of very high electric load (Na11, for example) and that of very low electric load (Cl17, for example), the former with 8 eV, the latter with 1.14 eV, filtering off the voltage excess beyond the limit allowed, avoiding damage or deterioration in the contacts of valences that participate in the chemical reactions,

5) - Select or reduce the electric tension to be applied in the next unit, thus not allowing it to be subject to a voltage exceeding its capacity,

6) - Allow, by means of series-equivalent and parallel-equivalent bindings, to reduce or increase, respectively, the total capacitance of the series of the electronic shell taken into consideration,

7) - Allow deuterotomes to be created with various capacitance values, as is the case of isotopes, otherwise there would be an idleness in the physical and chemical properties of the elements,

8) - Allow the existence of capacitors with variable electric loads in the deuterons that make up the deuterotomes of short or long structures, post-noble gas,

9) - Enable, upon the entrance of dielectrics (neutrons) in the deuterotome structure, the total binding energy to be increased and the binding energy to be decreased between the nucleons, thus allowing for the existence of different physical and chemical properties among deuterotomes,

10) - Allow capacitors to be formed with high capacitance by means of the physical volume-reduction phenomenon for noble gases,

11) - Allow capacitors with high capacitance to be able to carry out the function of the genetic pole for the synthesis of the deuterotomes and create the phylogenetic lineage for the capture of alpha particles,

12) - Allow the genetic poles–with high capacitance in keeping with the deuterotomic numbers–to guide the synthesis of deuterotomes with a long post-noble gas structure, as is the case of uranium (U92),

13) - Allow the formation of the chemical pole, opposed to the genetic pole, where the chemical reactions might occur by means of the valences available and consequent formation of molecules,

14) - Create a structure by deuterons juxtaposition that enables the orbital electrons to be so positioned as to face the interior of a round-shaped spirally set structure, the more characteristically spiraled the longest the structure composed by post-noble gas deuterons, as is the case of elements with deuterotomic numbers close to that of radon (Rn86) and uranium (U92),

15) - Allow the creation of the "electronic furnace", composed by 32 electrons, in the deuterotomes with deuterotomic numbers above those of astatine (At85), run by parallel-equivalent bindings,

16) - Allow capacitors (deuterons) to have structures that enable series-equivalent and parallel equivalent bindings to decrease or increase their capacitances, respectively,

17) - Allow capacitors (deuterons) to be able to capture more dielectrics (neutrons), in order to increase the mass number and decrease their capacitance, thus creating heavier isotopes,

18) - Allow, upon the emission of beta particles (b) starting from the neutron, the unit to achieve the features of a proton, there being an increase in the deuterotomic number and the creation of a new element,

19) - Allow a self-adjusting mechanism of capacitance to exist in deuterotomes with electric loads in limits not allowed, by means of emission beta particles (β), as is case of the elements of radioactive families of radium, actinium, uranium-235 and thorium,

20) - Allow orbital electrons to be found inside the iron deuterotome (Fe26/58) which conditioned the structure to enable an intra-deuterotomic solenoid-circulation to be created producing a permanent magnetic field once initiated,

21) - Make it possible, through the creation of the "electronic furnace" in intra-deuterotomic position, to protect the radioactive property against external agents, such as pressure, heat, electric loads, magnetic fields, to name but a few,

22) - Allow the "electronic furnace" created close to the deuterotome's genetic pole to generate a potential difference (ΔV) equal to 32 eV and the consequent emission of electrons against the wall opposite the genetic pole, thus producing collisions and vibrations and causing ruptures in the junctions between a pair of deuterium and the rest of the structure, by emitting alpha particles,

23) - Allow the neutron (dielectric) to become sensitized in inadequate levels of capacitance and electric tension, possibly participating in the stimuli of beta particles (β) emissions, giving it the features of a proton,

24) - Allow capacitors in series-equivalent bindings, upon acquiring the status of a noble gas, to undergo the physical volume-reduction phenomenon, as is the case of the deuterotome of helium (He2), neon (Ne10), argon (Ar18), krypton (Kr36) and xenon (Xe54).

25) - Allow capacitors (deuterons) in parallel equivalent bindings to be able to pass by noble gas radon (Rn86), without causing the latter to undergo the physical volume-reduction phenomenon and make it possible to create the "electronic furnace." (Radon did not become a genetic pole to the deuterotomes, which in deuterotomic numbers are ahead of them such as francium (Fr87), radium (Ra88) etc., as would be expected should it undergo the physical volume-reduction phenomenon seen in other noble gases).

2.6 – Capacitance of the atoms and deuterotomes

The capacitive balance of a deuterotome is vital for the integrity of the units that make it up. The mechanism that balances the capacitance within suitable limits can lie on the emission of beta particles (β). The growth in mass number through the admission of a dielectric (neutron) causes the capacitance to decrease, but the emission of beta particles (β) raises the Z deuterotomic number, reduces the number of dielectrics, increases the number of plates and the capacitance value. The mechanism that balances the deuterotome's capacitance value makes it possible to increase or decrease it as necessary also regulating the greatness of the tension to which it will be subjected. Should the tension applied through its plates be too much excessive, the dielectric might not be able to withstand the electric pressure and an arch would break out between the plates, and destroy the dielectric; the capacitor would go through a short circuit and would produce a flow of electric current of inadequate intensity.

Each capacitor is provided with characteristics of electric insulation against tensions that cannot be exceeded. An industrial capacitor needs a stand-by insulation tension in excess of at least half its load value. Thus, an industrial capacitor designed for 110 V should have a stand-by tension of 50 percent and be then capable of withstanding a tension of at least up to 170 V which can reach 250 VAC in a capacitor provided with ± 10 mF capacitance value.

The **work tension** of the capacitor is the maximum tension that can be applied without any danger of an arch being formed. This depends on the type of the dielectric material (k) and the dielectric thickness (d). As regards atomic capacitors, the type of the dielectric material will always be the same because the neutron composition remains unchanged, although the distance between the plates can undergo a variation through the entrance or exit of neutron, as well as the neutron transformation into a proton. In the following example the capacitance calculation shows that these values decrease upon the capture of a neutron (dielectric) and increase upon the emission of beta particles (β), thus raising the number of protons (plate): U92/238 + n——>U92/239 -β——> Np93/239.

By applying the equation to the **C** capacitance the following values expressed in eF are obtained: for U92/238, 0.9898; for U92/239, 0.9830; for Np93/239, 1.0005.

It is particularly difficult to calculate the capacitance of deuterotomes because the variables can be far too many. Several conditions can be mentioned that underline ongoing difficulties, namely:

1) The genetic pole, composed of a noble gas, is a multiplate capacitor that has gone through the physical volume-reduction phenomenon after it has become electrically neutral. This physical phenomenon has reduced the interplate distances and increased the capacitance. The reduction percentage is variable for each noble gas and different for neon and krypton, whose final values is twice the volume of hydrogen, while the other noble gases turn out to have a volume similar to that of the hydrogen atom,

2) In deuterotomes with valences 1, 2, 3 and 4, respectively the cases of sodium (Na11), magnesium (Mg12), aluminum (Al13) and carbon (C6), the deuterons (capacitors) that make up the post-noble gas structure are neither series- nor parallel-equivalent connected,

3) Deuterotomes with five or more post-noble gas deuterons will have series-equivalent arrangements should they have a deuterotomic number up to 54 (Xe54),

4) Single capacitors in series-equivalent bindings have a capacitance lower than the lowest capacitance of the capacitor in the series; the position of neutrons (dielectrics) situated between protons (plates) alters the capacitance value of the unit, upwards or downwards,

5) Single capacitors with parallel-equivalent bindings have the total capacitance equal to the algebraic sum of each capacitor,

6) The number of dielectrics (neutrons) between the plates (protons) in a multiplate capacitor with three plates and four dielectrics in the interval can have the dielectrics in different positions thus altering the capacitance values. A difference in capacitance will occur if the deuterons are series- or parallel-equivalent connected. The scheme below points out these possibilities.

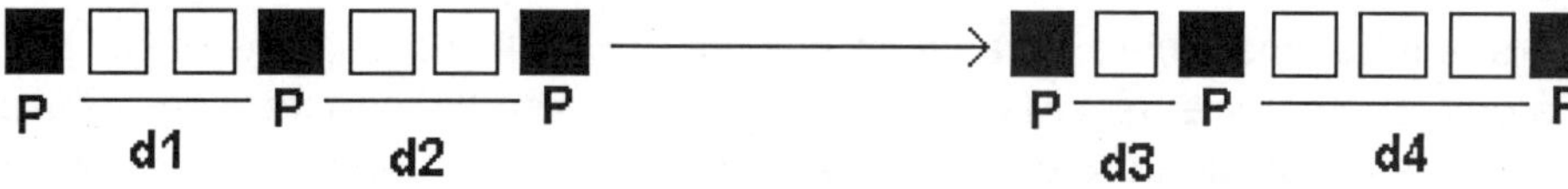

Fig. 15 - Variations of distance **d** with different number of interplate neutrons.

Thus, **on the left** scheme (Fig.15), the distance between the d1 and d2 plates is the same, because in the interval there are two dielectrics between each plate P. **On the right** scheme the d3 distance is shorter than that of d4 due to a change in the position of the neutron. These two variables alone (interplate distance and type of bindings of capacitors) make it significantly hard to calculate the capacitance.

7) The electric constant of neutron is currently a value taken at random, equal to 1 but this value is also the same for air and vacuum. Value **d** is equal to 1 and taken ad random too, but takes into account that the proton and the neutron alike have identical diameters. Thus, the A value represents the area of the hemisphere and is therefore, a constant value (1.5708).

After the synthesis of each noble gas has been completed their individual capacitance displays a certain value, which is altered upon going through the physical volume-reduction phenomenon. Because they are multiplate capacitors the volume-reduction phenomenon shortens the interval of the interplate distance, but maintains the same value for the plates area. The volume-reduction of the deuterotome of noble gas shortens distance **d** and increases the **C** capacitance. The table below illustrates the noble gas volume (compared to the volume of light hydrogen) after going through the physical volume-reduction phenomenon, as well as the density for each noble gas (expressed in g/cm^3) and their individual capacitance (expressed in eF).

Noble gases	Final volume of the noble gas after the volume-reduction	Density of the noble gas after the volume-reduction	Capacitance
He2/4	1	4	3.1416
Ne10/20	2	10	7.8540
Ar18/36	2	18	14.1372
Kr36/78	1	78	56.5488
Xe54/124	1	124	84.8232
Rn86/218	1	218	135.0888

The table shows that the capacitance values of noble gases are very high when compared to that of uranium (U92) that is 0.9898eF. Whereas noble gases are genetic poles for deuterotomes formed from them, they must carry out the functions of capacitors themselves, namely: store electric load, filter off any excess of load tension, stop the flow of inadequate electrical current and avoid spark from forming on the contacts of valences during chemical reactions.

In transmutations where a reaction consists of both the capture of an artificially accelerated alpha particle used as an atomic projectile and the spontaneous emission of neutron, the resulting deuterotomes will always have capacitance values higher than those of the deuterotomes they stemmed from. In Chadwick's bombardment (James Chadwick, English Physicist) in 1932 by making use of alpha particles against the deuterotomes of berilium-9 (Be4/9) the isotope of carbon (C6/13) was obtained which emitted neutrons before long. The corresponding equations are as follows:

(α, n) Be4/9————————>C6/13————————>C6/12 or
Be4/9 + α 2/4————>C6/13————————>C6/12 + n

The capacitance of each deuterotome formed in keeping with this reaction is as follows: Be4/9 = 1.2624eF; C6/13 = 1.3464eF; C6/12 = 1.5708eF therefore, increasing upon the capture of alpha particles and the emission of neutrons.

Another transmutation process, of low yield (one deuterotome transmuted with every one million alpha particles thrown against the deuterotome) is that of nitrogen-14 that is transmuted into fluorine-18 and oxigen-17 before long, in accordance with the following equation:

(α,π) N7/14————>F9/18———>O8/17 or
N7/14 + α——>F9/18———>O8/17 + π

The capacitance of the deuterotome that does not have any excess of neutrons resulting from the capture of alpha particles remains unchanged but decreases after the emission of protons (π), presenting the following values: N7/14 = 1.5708eF; F9/18 = 1.5708eF; 08/17 = 1.3962eF.

2.7 – Deuterium as single capacitor of deuterotomes and binding types

The deuterotomes formed from the genetic pole composed of xenon (Xe54), that is, from cesium (Cs55) onwards will make up a capacitive system with parallel equivalent bindings among themselves one by one, maintaining one to five orbital electrons in every last respective deuterons that make them up. The scheme below makes it possible to pinpoint the phenomena that can happen to electron.

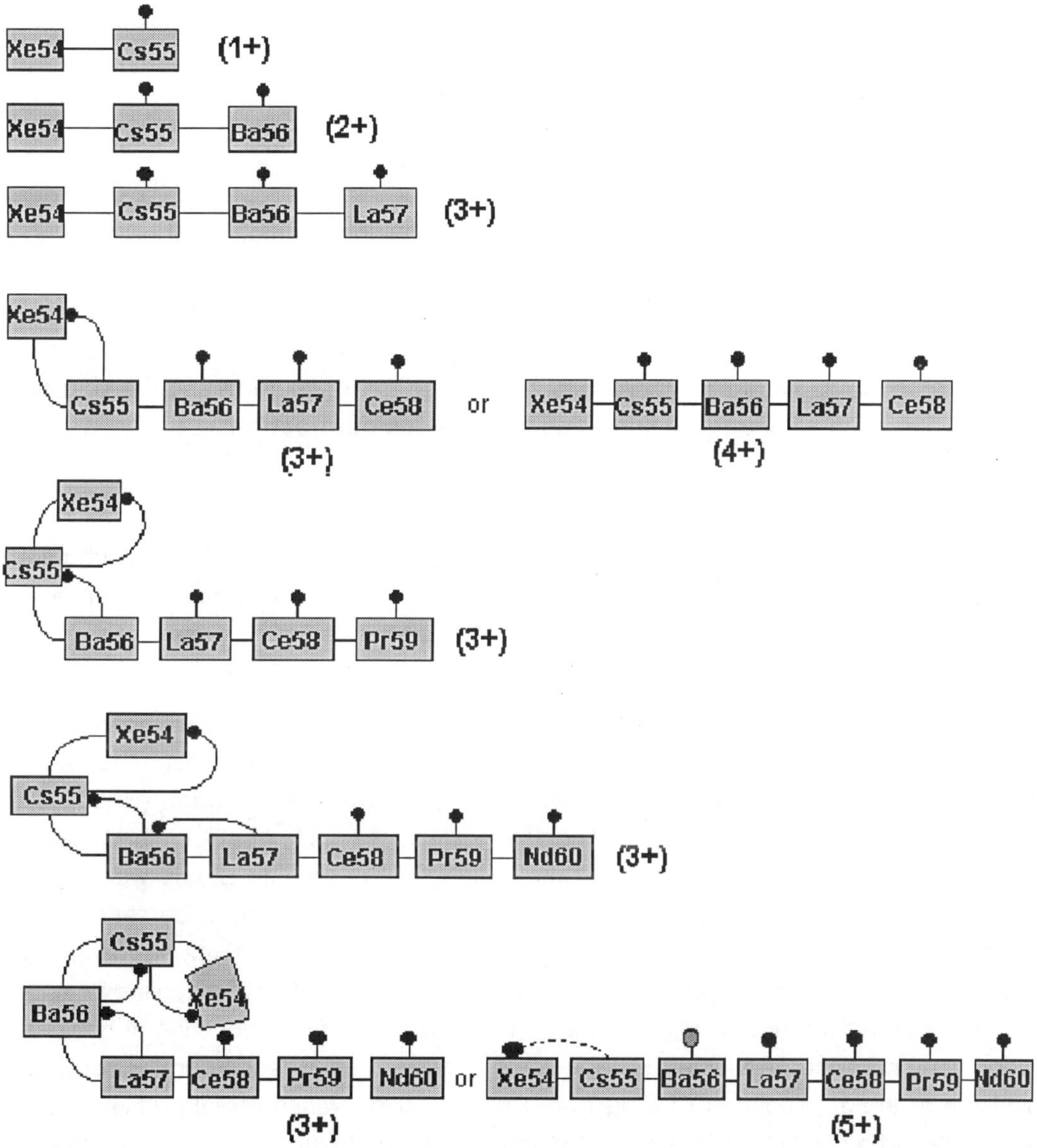

From scheme 1 to 3, the increment of deuterons results in an increase in the deuterotomic number (Z) and the number of free valences of high electric load (positive). Starting from scheme 4, cerium (Ce58) can react chemically with valences 3+ or 4+. If it remains without an electron binding, its valence will then be 4+, but if Cs55 electron is connected to the mass of the genetic pole composed of xenon (Xe54), cerium will act with valence 3+. That binding causes that Cs55 to be attracted towards Xe54, thus setting out the entanglement process of the deuterotome which will be more and more typical as the Z deuterotomic number increases. In scheme 5, praseodymium (Pr59) already has two parallel equivalent bindings showing the effect of attraction forces, reeling the deuterotome with bindings Xe54/Cs55 and Cs55/Ba56. In scheme 7, neodymium (Nd60) presents free valences 3+ and 5+. The bindings Cs55/Ba56 and Ba56/La57 of the deuterons (capacitors) can be undone as a result of chemical reactions under the action of pressure and/or temperature changes, shifting from three to five free valences.

Starting from samarium (Sm62) to lutetium (Lu71) the entanglement process upon the increase of Z maintains the number of free valences at 3+. The volume of deuterotomes should increase progressively but the entanglement process resulting from parallel equivalent bindings with the traction of the deuterium over the other, thus producing a deuterotome of spiraled form, prevents that increase from occurring and the volume is said to go through a contraction known as **lanthanides contraction**.

According to the prevailing atomic theory, that contraction would exclusively be due to an increase in the number of electrons in the N-shell or energy level 4s, where the number of electrons increases progressively from 18 to 32, without altering them in the 0-shell or energy level 5s, composed of nine electrons.

The parallel-equivalent bindings continue to be formed progressively producing a spiraled form as far as the deuterotome of uranium (U92) failing to go through the volume-reduction process upon reaching the deuterotomic number 86 of radon (Rn86). This physical phenomenon does not occur when the position of a noble gas is reached because the bindings are parallel equivalent. If the bindings of deuterium atoms were series-equivalent, the physical volume-reduction phenomenon would then be repeated as happens to deuterotomes with genetic poles composed of helium (He2), neon (Ne10), argon (Ar18) and krypton (Kr36). If there were a volumetric reduction of radon the formation of the “electronic furnace” would not occur. Thus, as the volume-reduction of noble gas radon does not occur, francium (Fr87) when synthesized undoes the parallel equivalent bindings of the 32 electrons ranging from Cs55 to Rn86, which remain offset and are aggregated at a point close to the genetic pole that will make up the “electronic furnace.” The electrons of free valences (6+) of the deuterotome of uranium (U92) are not part of the electrons that make up the “furnace” which, in order to have valences 3+ and 4+, need to have three or two electrons connected in parallel equivalent bindings without making a part of the “furnace”.

The following scheme represents in (a) chromium deuterotome (Cr24) with capacitive bindings of the series-equivalent type linked two by two; in (b) neodymium deuterotome is represented (Nd60) with capacitive parallel equivalent bindings.

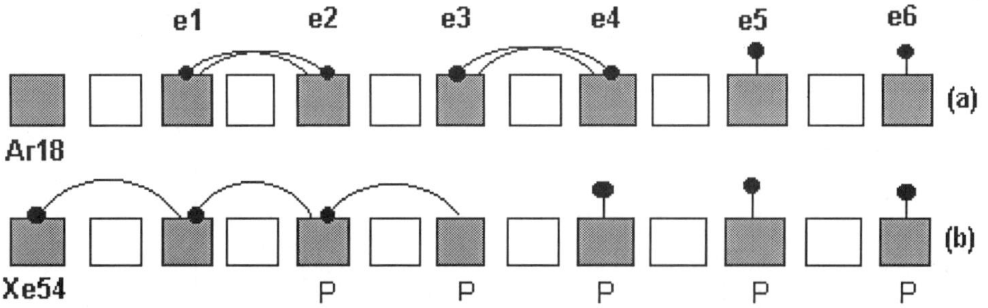

Fig 16. Scheme showing the capacitive bindings in **series-equivalent** in (a) and parallel **equivalent** in (b). In (a) the genetic pole is composed of noble gas argon (Ar18) and in (b) xenon (Xe54). The black squares represent protons and the white ones neutrons. In scheme (a) the e5 and e6 electrons are not connected representing free valences (2+); in scheme (b) the e4, e5 and e6 electrons are not connected and represent free valences (3+).

The radioactive families of heavy deuterotomes have elements with deuterons (capacitors) in parallel equivalent bindings. During the processes of radioactive degradation the capacitance of each element formed from uranium (U92) up to lead (Pb82) or from thorium (Th90) up to lead undergo a progressive upward variation. The emissions of alpha particles cause the capacitance values to decrease, while the emissions of beta particles (β) increase them. The final element, stable (Pb82/206, 207 and 208), has a capacitance higher than the elements from which they originate: U92/238, U92/235 and Th90/232, respectively. The following charts show the variations in capacitance values (vertical line) in relation to the number of neutrons in the "alpha-appendix" (horizontal line). However, when the value of capacitance total (Ct) of deuterotomes is calculated as per the equation for capacitors in series- and parallel connections, the results then vary, averaging at around $1.529.10^{-3}$ to 54.10^{-3} eF for series bindings (from Li3/6 to I53/129) and around 84.823 to 129.86 eF for parallel (from Cs55/133 to U92/238 of radium family). Up until now these equations have not been used due to the evident complexity resulting from the physical volume-reduction phenomenon of noble gases. But the capture of alpha particles and neutron as well as the beta emission somehow unbalance the capacitance values.

Based on the general equations of arrangements, the calculations point out that the capture of alpha particles reduces the capacitance of new deuterotomes then formed, when these arrangements are in series, and increases it when in parallel. The capture of two or more neutrons and the emission of a beta particle cause the capacitance of the newly formed deuterotome to decline if the arrangements are in series and make it increase when in parallel. The capture of a neutron and the emission of a beta particle increase the number of receptive plates (proton) by 1, but this is not considered a new capacitor for it lacks a neutron in order to complete it. Thus, a new capacitor will not be formed until one other capture of neutron occurs. That, besides others, is a crucial issue for calculations to be made, which is why the general equation for capacitance was currently adopted for analytical purposes without taking into account the arrangement type. Only the variation in the number of protons and neutrons is considered. Therefore, final basis leading to conclusive calculations are yet to be set up.

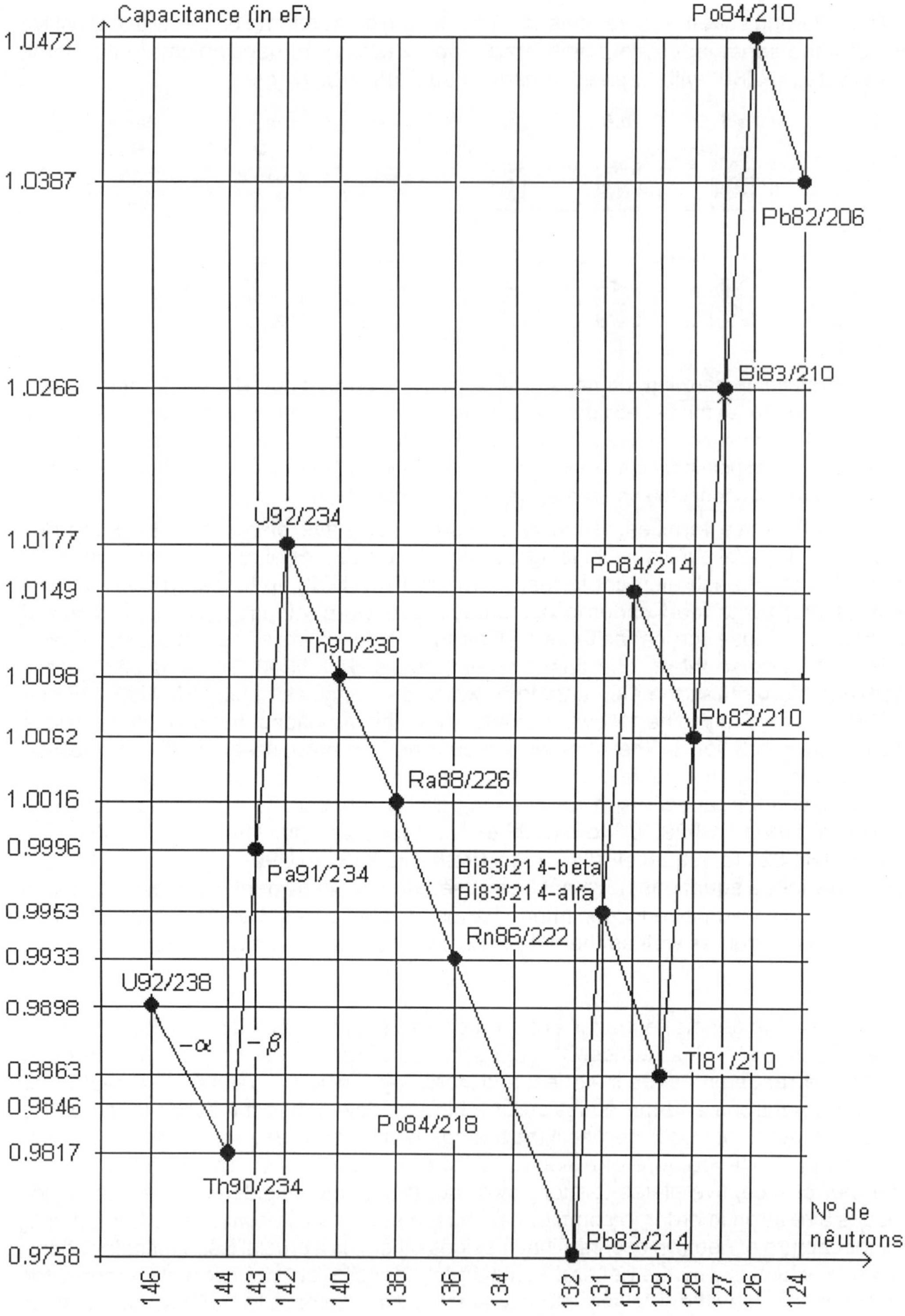

Fig. 17 – Radioactive degradation of radium family

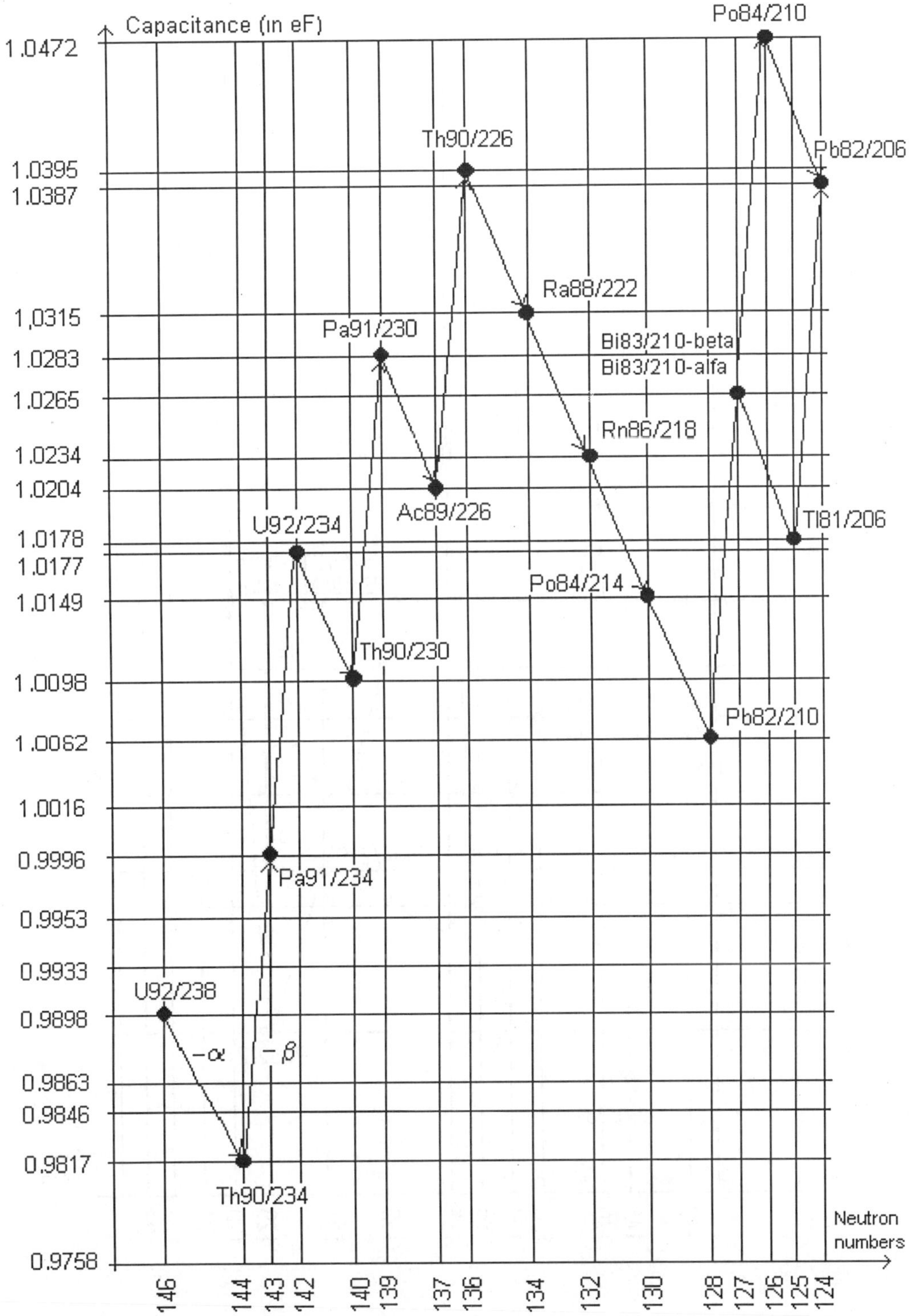

Fig. 18 – Radioactive degradation of actinium family

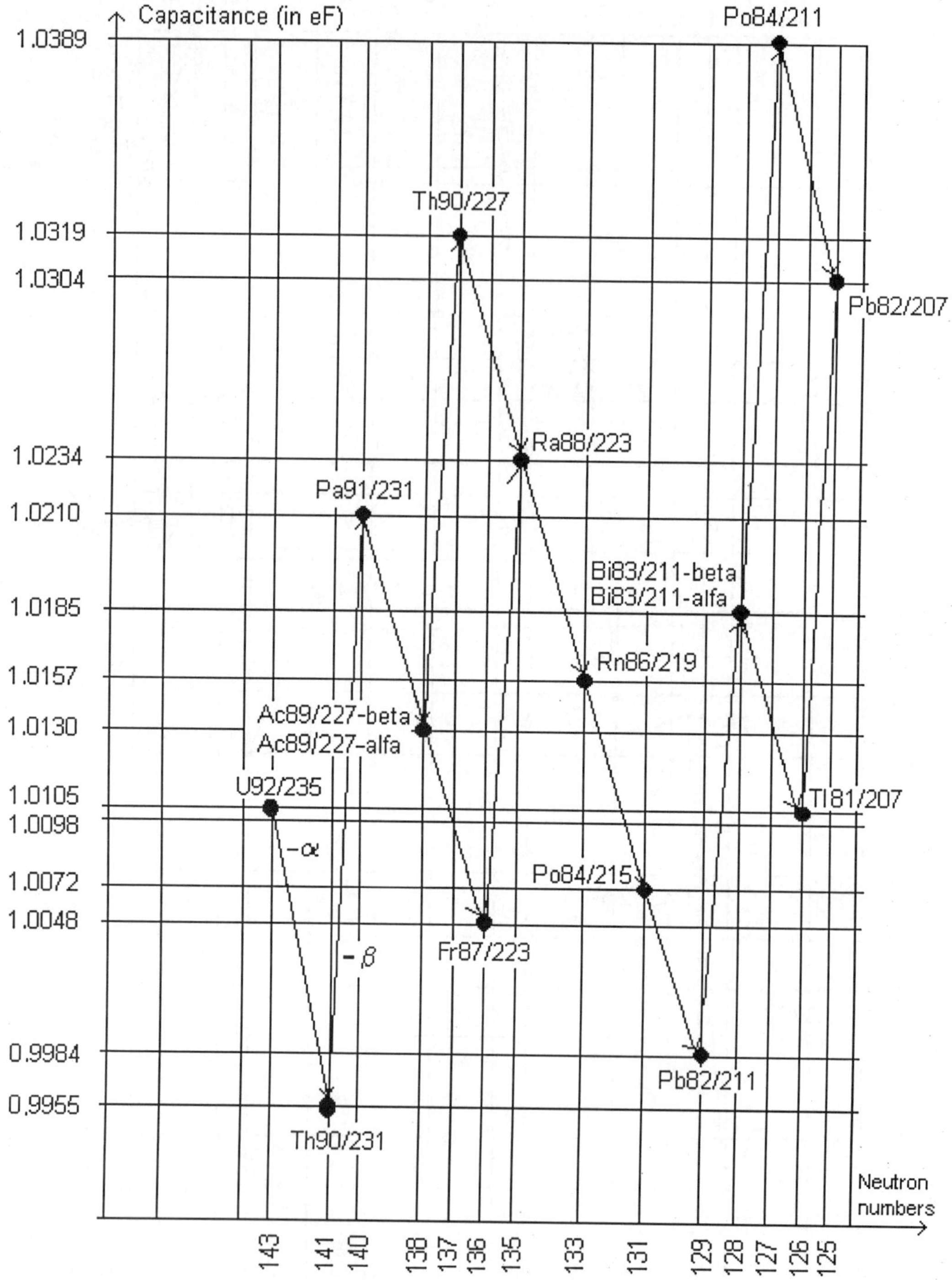

Fig. 19 – Radioactive degradation of U92/235 family

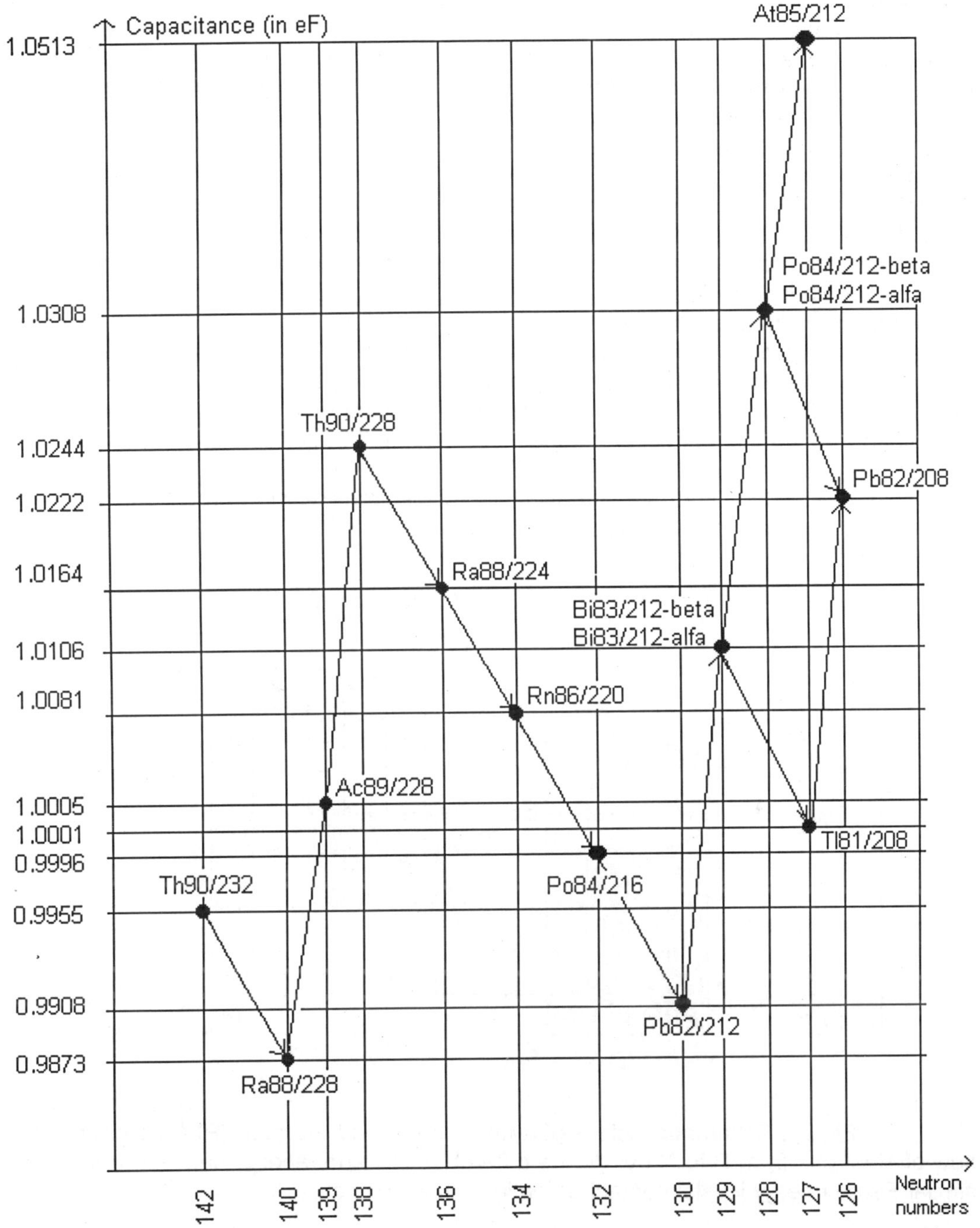

Fig. 20 - Radioactive degradation of thorium family

2.8 – Contact potential in Neutron/Proton junctions

What enables the contact between the surface of a proton with that of a neutron is the existence of a potential difference between these surfaces. The potential difference is related to the binding forces of the nucleons (Ben). The electrons are aligned on the proton surface that comes in contact with the neutron. The total energy, represented by the amount of electrons, is far greater in the proton's structure; it must have an energy level similar to the Fermi energy (E_F) of conductor metals. The proton and neutron have different levels of Fermi energy. The electrons close to the surface of the proton with high Fermi level can move on to occupy empty states well above the neutron's Fermi level. The potential difference is then settled down because both the proton and neutron consist of structures where the electrons are moving almost freely from the proton to the neutron. Electrons on the neutron's contact surfaces do not alter its features of an electrically neutral unit. The orbital electron, which neutralizes the neutron's load, does not participate in the flow of electron from the proton to the neutron. Thus, the neutron becomes a negatively loaded unit and the proton a positively loaded unit. A potential difference is developed through the contact and adhesion of the neutron to the proton, denominated **contact potential**. Figure 21 illustrates in (a) the scheme of a neutron (N) in contact with a proton (P) with different Fermi levels and a flow of electrons passing through the interstice of the lentonian plates of the proton to that of the neutron; in (b) the Fermi levels of both are well balanced.

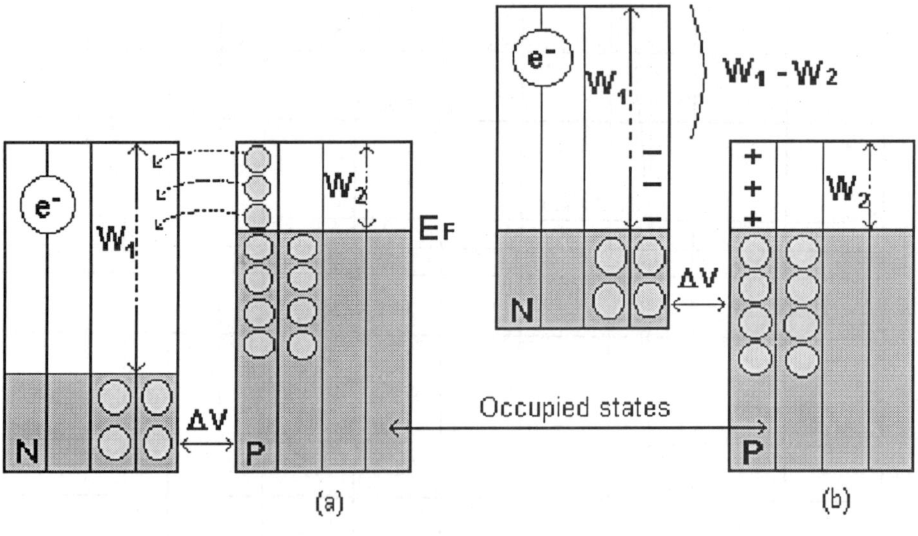

Fig 21. Schematic picture of a neutron (N) and a proton (P) **(a)** showing the flow of electrons from P to N due to the difference of Fermi energy between them. **(b)**: similar Fermi energy levels; contact potential V in N-P junction.

In comparison with conductor metals, if one designates by **W** the energy necessary to remove an electron from the highest Fermi level (of proton) to the lower Fermi level (of neutron), the contact potential V in Fig 21 will be figured out by the difference between the **work functions** (W1-W2) of the two structures divided by the energy of an electron **e**. Thus, the equation can be as follows:

$$V = \frac{W_1 - W_2}{e}$$

where **V** is the difference of contact potential, W1-W2 the difference between work functions of the neutron and proton, respectively, and **e** is the electron's electric load (e = 1.602 x 10^{-19} C).

The attraction force between the proton and neutron is dependent on the nucleon's binding energy. That energy is calculated per nucleon and expressed in MeV (mega-electron-volts). The highest binding value of energy per nucleon can be found in the deuterotome of iron (Fe26/56) and the lowest in uranium-238 (U92/238). For Fe26/56, that energy is equal to 8.791 MeV per nucleon; for uranium-238 equal to 7.571 MeV.

When an additional junction is considered with the admission of one more neutron to the neutron/proton unit already formed, there occurs a group consisting of two junctions formed by neutron/neutron/proton (N-N-P), which can be represented by the following scheme:

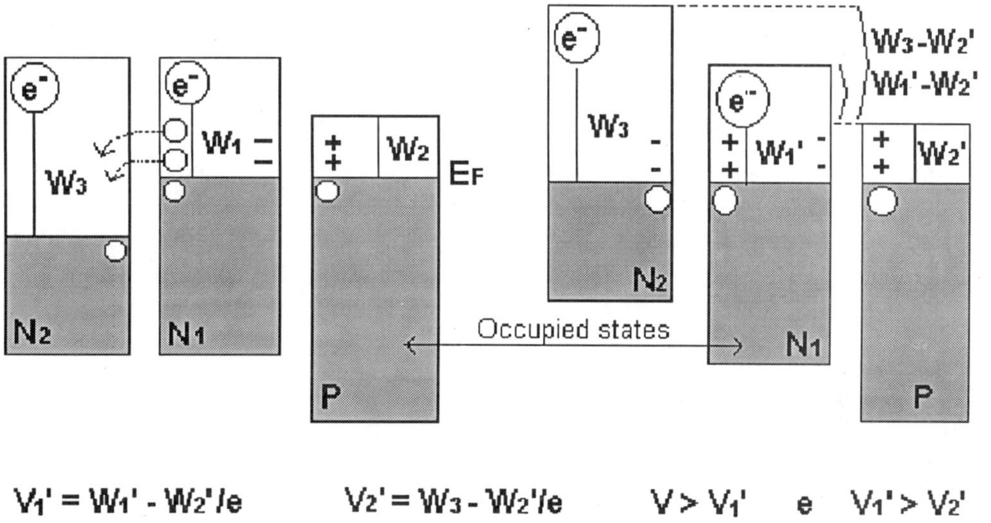

Fig. 22. Scheme showing two neutrons N, and one proton P with two junctions where they form a V2 and V1 contact potential with lower Fermi energy level (left). When the electron flow from N1 to N2 the Fermi level is balanced (right).

Should the contact potential V2 ' be lower than V1' and V1' in Fig. 22, than V in Fig. 21 then the binding energy per nucleon will decrease as the deuterotome acquires a larger number of neutrons in comparison with number of protons, which occurs to elements that become heavy where the Z value tends to 92, A (deuterotomic mass) to 238 and N (number of neutrons) to 146.

CHAPTER 3

The synthesis of matter

3.1 – Synthesis mechanisms

As soon as great masses of electrons and protons have been synthesized that had been clustered in atoms of light hydrogen, deuterium and tritium, the synthesis processes then begin. The fusion process of deuterium with tritium originates the first complex deuterotome, namely, unstable He2/5, capable of being disintegrated into alpha particles (α2/4), beta particles (β) and neutrons with the consequent production and release of thermal energy on a large scale. Eight possible mechanisms for the synthesis of deuterotomes can be then pointed out:

1 – Proton-proton cycles (protons fusion)
2 – Carbon cycle (protons capture)
3 – Fusion of light deuterotomes
4 – Fusion of deuterotomes with intermediate deuterotomic numbers
5 – Capture of alpha particles
6 – Capture of neutrons
7 – Capture of neutrons and emission of beta particles (β)
8 – Emission of neutrons

Before focusing on the details of the basic composition of deuterotomes, one can point out that the synthesis of matter has already occurred and still occurs at a stellar temperature with the initial fusion of deuterium with tritium and the formation of a transitory and unstable helium He2/5 deuterotome, which is disintegrated into neutrons, alpha particles (α) and beta particles (β) with an intense release of thermal energy. When no more hydrogen atoms are available, the fusion process is extinguished forthwith and as the temperature drops the expanded matter is contracted. After the volumetric reduction of matter has occurred, the process of radioactive degradation of heavy and instable deuterotomes begins as in U92/238 whose deuterian structures between Xe54 and Rn86 would not have their electrons neutralized or even undergone the physical volume-reduction phenomenon, thus forming the agglomerate of intra-nuclear electrons ("electronic furnace"), setting off a number of conditions for nuclear instability and radioactive property, that is, the release of pairs of deuterium into their structure–which make up alpha particles–and the release of beta particles (β) and gamma rays (γ), decreasing the deuterotomic number and the mass number of thallium (T181/230) which becomes stabilized in lead (Pb82/206, 207 and 208).

The radioactive degradation of uranium (U92) starts off after matter has cooled down at an astral temperature. Because there are radioactive deuterotomes like thorium (Th90) whose half-life is 25 billion years (25.10^9), a time estimate can be figured out as to the period elapsed from the time the reactions of nuclear fusion ceased to occur up to now.

3.2 – Constitution of deuterotomes

An analysis on the number of deuterons that successively participate in the synthesis of elements (deuterotomes) from the most simple to the most complex and heavy ones, highlights a phylogenetic lineage of deuterotomes from lithium (Li3) to uranium (U92), the analysis of which will be brought up later on.

The new deuterotomic theory, as a proposal, fits an assumption as to how matter is formed from lentons and the concept that the atoms of hydrogen made up from lentons were composed in the cornerstone of the theory. Light hydrogen (H1/1), which consists of only a proton and an orbital electron juxtaposed to the proton situated in the lower K-shell or energy level s1 in an self-neutralization process, is made up of a neutron (n) devoid of electric load and atomic mass equal to 1. A neutron interposing a proton plus an electron originated the deuterium atom (D1/2) and the tritium atom (T1/3) with the presence of two neutrons. Provided with fundamental characteristics of a capacitor, both the deuterium and the tritium atoms have become the so-called raw material for the synthesis of elements, as it were.

The number of deuterons that can be found following a noble gas in a given string makes it possible to figure out the volume of the deuterotome then formed. Oxygen, for example, has six deuterons following noble gas helium (He2/4) that composes the deuterotome's genetic pole whose composition is as follows: He2/4 + 6 D1/2 = O8/16. The following illustration refers to the example (Fig. 23).

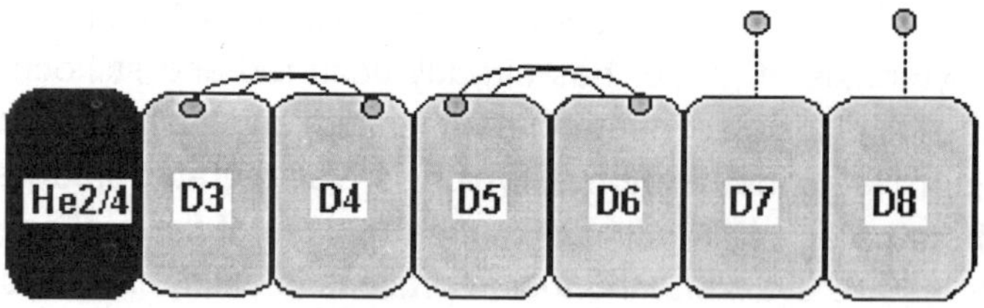

Oxygen (O8) deuterotome can also be illustrated in bi-dimensional form

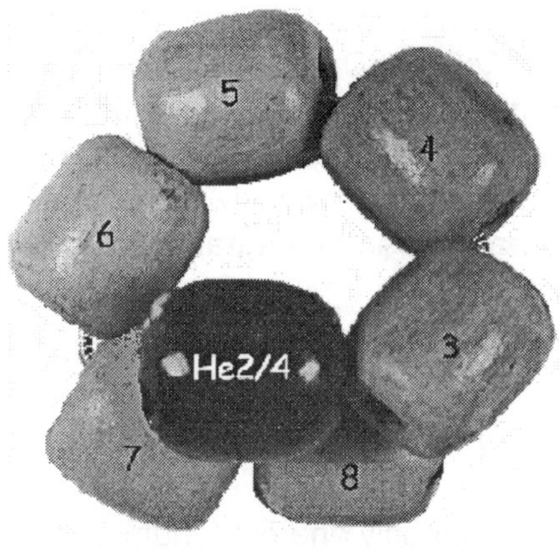

Fig. 23 – Illustration in ring-like and elliptical shapes of oxygen (08) deuterotome composed of noble gas helium (He2/4) in the genetic pole and six deuterons (D3, D4, D5, D6, D7 and D8). Deuterons D7 and D8 make up the chemical pole of that deuterotome that has thus two free valences with 1.33 eV emf each. Likewise, the deuterium atoms D3/D4 and D5/D6 are neutralized by bindings among them, not represented in this Figure.

The deuterotome of carbon (C6) is composed of noble gas helium (He2/4) in the genetic pole followed by four deuterium atoms; it has three isotopes (C6/12, C6/13 and C6/14). The arrangements in ring-like and elliptical shapes of this deuterotome, places it in the center of a tetrahedron. The photo below highlights that feature which is the aspect of carbon itself.

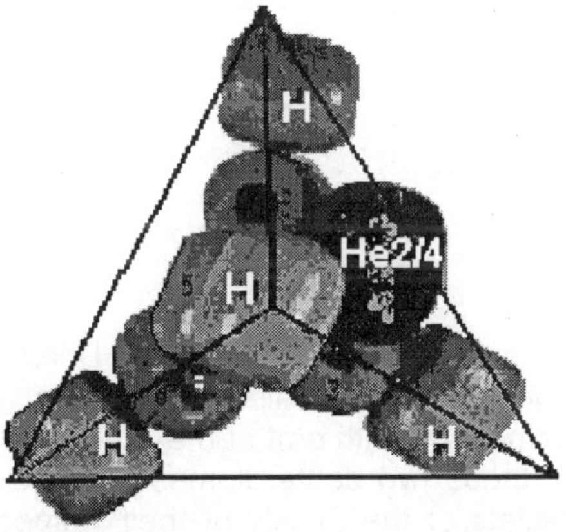

Fig. 24 - Ring-like and elliptical illustration of carbon (C6) deuterotome combined with four atoms of hydrogen, making up methane (CH4) structure. The carbon position is highlighted inside the thetrahedron. The genetic pole is formed by helium (He2/4) while the chemical pole consists of the four post-noble gas deuteriuns, namely D3, D4, D5, and D6 each havinq a free valence and 2 eV emf in each deuterium.

Elements following up after noble gas radon (Rn86) produce physical evidence that they have been synthesized up to uranium (U92) due to their radioactive properties without there being no physical volume-reduction phenomenon undergone by the said noble gas. The composition of the deuterotome of uranium (U92) is therefore as follows: Xe54 + 38D = U92. In uranium's structure, radon does not function as a genetic pole. In the composition of the "electronic furnace" of uranium, 32 orbital electrons are present, a condition that exists only in heavy radioactive deuterotomes consisting of Xe54 + 38D or in elements that take part in radioactive families. Therefore, the existence of U92/238 with a structure composed of Rn86 + 6D is not very likely, but theoretically its existence is not ruled out; if it does exist its percentage will be far too minimal and will not consist of a radioactive deuterotome like all the others, for the "electronic furnace" will not be available nor will it have a phylogenetic code previously established to release alpha and beta particles in a process of natural radioactive degradation. In addition, the "alpha-appendix" in its chemical pole will be absent.

The noble gas that forms the genetic pole of a deuterotome embodies the code for the synthesis of the next deuterotomic series. A certain series sets out with a noble gas (He2, Ne10, Ar18, Kr36, Xe54) and its synthesis goes on until the position of a halogen (F9, Cl17, Br35, I53) is reached.

There are deuterotomes with only one deuterium after the noble gas as is the case of lithium (Li3), sodium (Na11), potassium (K19) etc.; others with intermediate number of deuterium atoms for example, beryllium (B5), titanium (Ti22) and others with a large number of deuteriums, as that of bromine (Br35), iodine (I53) etc. These examples can be seen in the figure below:

Fig. 25 – Illustration of the deuterotome of boron (B5) and lithium (Li3). Boron has deuterons D3, D4 and D5 with emf of 2.66 eV each, all with its valences (3+) available. Lithium has its only deuterium post-noble gas, namely D3 with emf of 8 eV, also with its only valence (1+) available. The genetic pole of these two deuterotomes is noble gas helium (He2/4). The chemical pole of both consists of the totality of their respective deuterons because they have a short structure.

Deuterons D15 and D16 of chlorine's deuterotome (Cl17) can release their valences transforming that element into trivalent (3^-); deuterons D13 and D14 can also release their valences and the element can change into pentavalent (5^-). Likewise, the release of the valences of deuterons D11 and D12 causes chlorine's deuterotome to be heptavalent (7^-), each deuteron always having an emf of 1.14 eV, not losing its receptive valence (negative and with low energy chemical reaction). Fluorine, however, in all of its chemical molecular compositions always appears as negative monovalent: potassium fluoride (KF), fluoride acid (H2F2), sulfur VI fluoride (SF6) and sulfur V fluoride (S2F10).

The capture of alpha particles increases the deuterotome number two by two and the mass number four by four in the series L, M, N, O-shells until the noble gas at the end of each series is reached. If the synthesis of the noble gas takes place, the entire series then has its deuterons with 1 eV emf, which characterizes the total neutralization in each series, up to xenon (Xe54). Upon neutralization, the physical volume-reduction phenomenon acts by reducing the volume of each one to the volume of the hydrogen atom, except for noble gases neon and argon, whose volume are twice that of hydrogen atom (proton).

In spite of the fact that two deuterotomes have any identical number of post-noble gas deuterons, e.g., carbon (He2/4 + 4D) and silicium (Ne10 + 4D), the volume of noble gases in each one is not identical and neither are their respective capacitances.

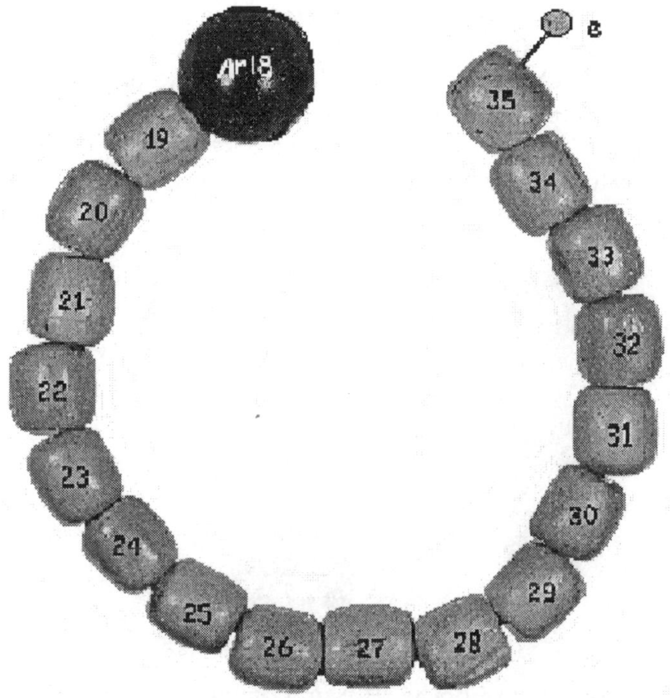

Fig. 26. Illustration of bromine deuterotome (Br35) with its genetic pole constituted by noble gas argon (Ar18), which has double volume. The deuterons, which range from D19 to D35, compose the structure of the deuterotome, here represented lineally. Deuterium D35 makes up the deuterotome scheme, here represented on a linear basis. Deuterium D35 is the chemical pole of that monovalent (1^-) deuterotome, with very low emf of 1.05 eV. The electrons of each deuterium atom are not represented in the figure, but they are connected in pairs, in series-equivalent bindings.

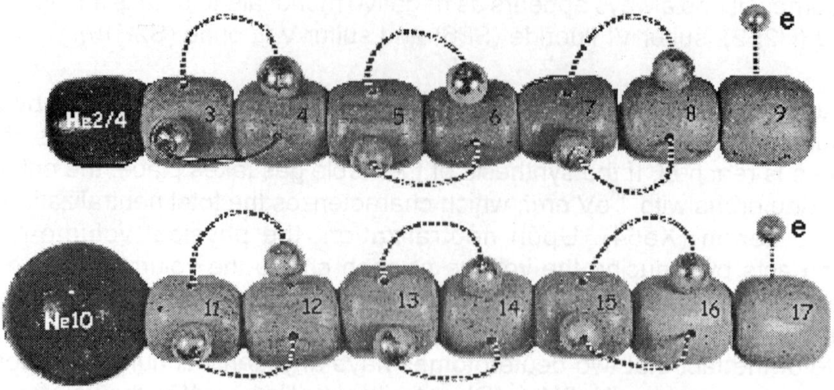

Fig. 27 – Illustration of fluorine (F9) and chlorine (Cl17) deuterotomes. The genetic pole of the former is formed by noble gas helium (He2/4)–top of photo–and the latter, by neon (Ne10). See the volume difference in the noble gas that composes the genetic pole of each deuterotome. Each of the two deuterotomes has seven deuterium atoms after the noble gas–D3 to D9 for fluorine and D11 to 17 for chlorine–both with very low electromotive force of 1.14 eV each, monovalent (one free valence in the deuterons D9 and D17) in the illustration above.

The illustrative picture below demonstrates the composition of the deuterotomes with three deuterons, with their respective noble gases that compose the genetic pole:

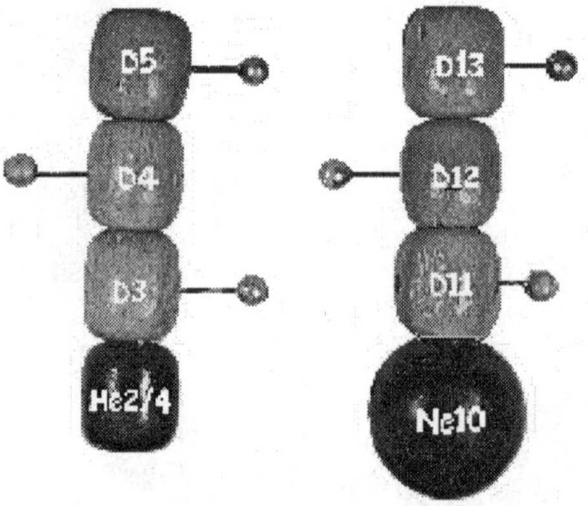

Fig. 28. – Illustration of boron (B5) and aluminum (Al13) deuterotomes. The genetic pole of the former (left of photo) is formed by helium (He2/4) and latter by neon (Ne10), which has a double volume. Both are trivalent (3+) and have 2.66 eV emf in each valence. The chemical pole of each is represented by the totality of post-noble gas deuterons.

He2 (Helium)	Ar + 5D = V23	Kr + 10D = Pd46	Xe + 15D = Tm69
He + 1D = Li3	Ar + 6D = Cr24	Kr + 11D = Ag47	Xe + 16D = Yb70
He + 2D = Be4	Ar + 7D = Mn25	Kr + 12D = Cd48	Xe + 17D = Lu71
He + 3D = B5	Ar + 8D = Fe26	Kr + 13D = In49	Xe + 18D = Ha72
He + 4D = C6	Ar + 9D = Co27	Kr + 14D = Sn50	Xe + 19D = Ta73

He + 5D = N7
He + 6D = O8
He + 7D = F9
He + 8D = Ne10

Ne10 (Neon)
Ne + 1D = Na11
Ne + 2D = Mg12
Ne + 3D = Al13
Ne + 4D = Si14
Ne + 5D = P15
Ne + 6D = S16
Ne + 7D = Cl17
Ne + 8D = Ar18

Ar18 (Argon)
Ar + 1D = K19
Ar + 2D = Ca20
Ar + 3D = Sc21
Ar + 4D = Ti22
Ar + 10D = Ni28
Ar + 11D = Cu29
Ar + 12D = Zn30
Ar + 13D = Ga31
Ar + 14D = Ge32
Ar + 15D = As33
Ar + 16D = Se34
Ar + 17D = Br35
Ar + 18D = Kr36

Kr36 (Krypton)
Kr + 1D = Rb37
Kr + 2D = Sr38
Kr + 3D = Y39
Kr + 4D = Zr40
Kr + 5D = Nb41
Kr + 6D = Mo42
Kr + 7D = Tc43
Kr + 8D = Ru44
Kr + 9D = Rh45
Kr + 15D = Sb51
Kr + 16D = Te52
Kr + 17D = I 53
Kr + 18D = Xe54

Xe54 (Xenon)
Xe + 1D = Cs55
Xe + 2D = Ba56
Xe + 3D = La57
Xe + 4D = Ce58
Xe + 5D = Pd59
Xe + 6D = Nd60
Xe + 7D = Pm61
Xe + 8D = Sm62
Xe + 9D = Eu63
Xe + 10D = Gd64
Xe + 11D = Tb65
Xe + 12D = Dy66
Xe + 13D = Ho67
Xe + 14D = Er68
Xe + 20D = W74
Xe + 21D = Re75
Xe + 22D = Os76
Xe + 23D = Ir 77
Xe + 24D = Pt78
Xe + 25D = Au79
Xe + 26D = Hg80
Xe + 27D = Tl 81
Xe + 28D = Pb82
Xe + 29D = Bi83
Xe + 30D = Po84
Xe + 31D = At85
Xe + 32D = Rn86
Xe + 33D = Fr 87
Xe + 34D = Ra88
Xe + 35D = Ac89
Xe + 36D = Th90
Xe + 37D = Pa91
Xe + 38D = U92

3.3 – Fundamental atomic particles for the synthesis processes

A) – Alpha particles (α2/4) are those with two positive electric loads having the nuclear structure of helium (He2/4) deuterotome that has lost its two electrons (experiment by Wilson and Milikan). Should an alpha particle receive two electrons, it becomes the deuterotome of helium (experiment by Ramsay and Soddy, Sir Rutherford and Royds). Alpha is a particle in movement whose initial speed is 20,000 km/s. Its power to penetrate matter is minimal and is limited to 7 cm in a layer of air or in a metal sheet 0.1 mm thick. It has, however, great ionizing power and is capable of turning gases encountered in its neighborhood, into electrical conductors, under normal pressure; an alpha particle ionizes 200,000 molecules of a gas.

During the synthesis process of matter, the capture of an alpha particle causes the receiving deuterotome to increase its deuterotomic number by two units and the mass number by four units, and the deuterotomic number of the receiving deuterotome is always an even one. Highlighted in the following examples are the three mechanisms of performance of each one of the three fundamental particles that lead to the deuterotome synthesis.

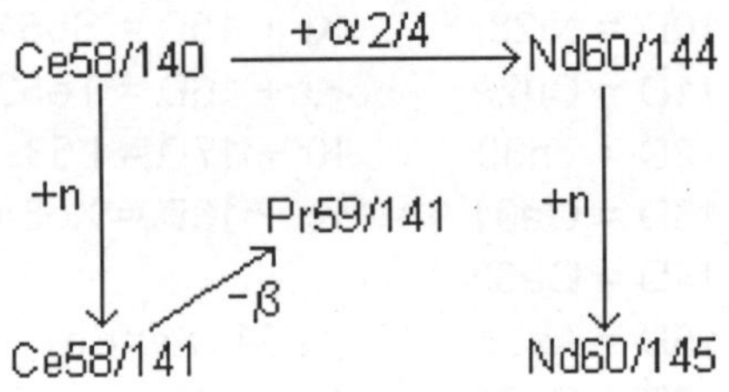

B) Beta particles (β)

is variable in keeping with their speed; for lower speeds, the mass averages 1/1,836 of the proton mass that forms the hydrogen atom; at a speed of 270,000 km/s, the electron mass has this value reduced by half. Beta rays (β) are similar to the cathode rays of the Crookes' tube, but they are provided with greater speeds–that is, with shorter wavelength.

According to the current atomic theory, beta rays (β) are regarded as essential electrons. In other words, part of the atom nucleus (forming the neutrons), whereas cathode rays, are composed of orbital electrons that gravitate around the nucleus making up the electronic crown.

Beta rays are particles with little ionizing property. Under conditions of normal pressure, in the air, an electron can only ionize five to seven molecules.

During the synthesis mechanism of matter, the emission of beta particle, by a radioactive deuterotome raises the deuterotomic number by 1, thus reducing the deuterium's electromotive force and increasing the number of capacitors by 1. In the following example: Cr24/55 – β

2.57 eV. The beta particle emitted by a radioactive deuterotome corresponds to an electron that gets loose from the neutron, conferring upon the latter a proton state.

C) Neutrons

1. They consist of an atom of light hydrogen (H1/1), whose orbital electron is juxtaposed to the proton and, consequently, neutralizing it. An essential particle for the synthesis of matter, they determine a rise in the mass number by 1 to the deuterotome that captured it. One neutron added to the deuterium atom–which acts as a capacitor–duplicates the space existing between the plates and consequently reduces the capacitance of the deuterotomic isotope of greater mass number. In the example ahead: Ti22/48 + n ——> Ti22/49, the capture of one neutron raises the mass number by 1 and reduces the

3.4 – Nuclear fusion as source of particles

It is common knowledge that all matter is transformed, basically, by nuclear fusion and radioactive degradation, the former releasing a huge amount of energy, often colossal.

The solar energy that results from the nuclear fusion between deuterium (D1/2) and tritium atoms (T1/3) is an example of an enormous source of thermal and photonic energy levels.

The synthesis of matter at stellar level should have basically and most likely occurred, via the eight processes cited below.

1) Proton-proton cycles,
2) Carbon cycle,
3) Fusion of light deuterotomes,

5) Capture of alpha particles,
6) Capture of neutrons,
7) Neutrons capture and emission of beta particles (β),
8) Neutrons emissions (according to the “Table of the possible pathways for synthesis of deuterotomes”).

The nuclear fusion of deuterium (D1/2) with tritium (T1/3) produces helium (He2/4) deuterotome, albeit indirectly, intermediate reactions are present which will be referred to later on. This fusion reaction is one of thermonuclear nature with the consequent production of thermal energy and emission of radiations. The nuclear fusion of these two heavy isotopes of hydrogen atoms must undergo the production of a deuterotome–isotope of helium–with a mass number which is helium-5 (He2/5), unstable and short-lived which disintegrates, emitting alpha (α2/4) and beta particles (β), gamma rays (γ) photons and neutrons, besides helium-5 (He2/5) deuterotomes and an isotope that is scarcely formed-

$^{-4}$%. The summarized equation

of nuclear fusion is symbolically shown below:

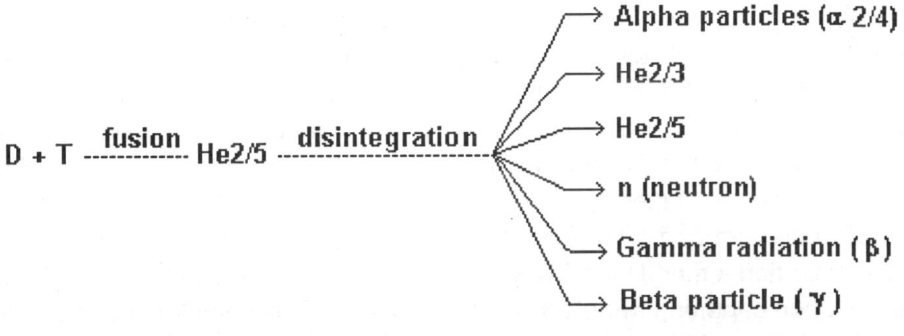

This synthesis process is continuous and everlasting, but tends to be extinguished when one of the two atoms decreases in concentration and shifts to a level below the limit necessary for the nuclear fusion to occur. However, the nuclear fusion can occur between two deuterium atoms (D + D) or between two tritium atoms (T + T), as thermonuclear reactions as well.

The alpha (α) and beta particles (β) are restored and originate helium-4 (He2/4) deuterotome. The isotope of mass number equal to 5 (He2/5) is short-lived and unstable, but can capture a neutron and give rise to helium-6 (He2/6), also transitory, which will emit a beta particle (β the category of "extinct isotopes", that is, they had a fleeting existence during the synthesis process, and are no longer found in nature. The isotope of helium-3 (He2/3), formed upon the disintegration of He2/5, captures an alpha particle, originating the deuterotome of berilium-7 (Be4/7), which will continue up to oxigen-15 (O8/15), passing by carbon-11 (C6/11). The deuterotome of helium-4 (He2/4), upon capturing an alpha particle, changes into berilium-8 (Be4/8); the latter also captures an alpha particle, and charges into carbon-12 (C6/12); this synthesis lineage continues up to calcium-40 (Ca20/40), where the synthesis of the lineage stops.

Not until the stage of carbon-12 synthesis (C6/12) is reached, can the synthesis by the carbon cycle mechanism be active, and, consequently, carbon-13 (C6/13), nitrogen-14 (N7/14), oxigen-15 (O8/15) and nitrogen-15 (N7/15) can be synthesized. The carbon cycle is really important because it does not deplete carbon-12 for its amount encountered at the end of the cycle is similar to that observed at the beginning; a whole cycle captures four protons (P) from the matter mass where the synthesis is processed, this cycle emits positrons or positive electrons (e+), neutrinos (v) and photons (quanta).

The atoms of deuterium and tritium, as atomic capacitors, play on important role towards the synthesis of elements. The different aspects between deuterium and tritium–they possess one and two dielectrics, respectively–offer a possible basic condition for the onset of the synthesis of deuterotomes to begin from the fusion between these two atoms. In the atomic or nuclear fusion, the deuterotome of helium-5 (He2/5) is an unstable isotope that disintegrates quickly, separating the units that compose it. The nuclear fusion and the subsequent disintegration seem to be the main source capable of producing a great amount of particles necessary for to the synthesis of the deuterotomes.

As capacitors, the atoms of deuterium and tritium, arranged in series-equivalent or parallel equivalent bindings highly concentrated and in a critical atmosphere, can supply an electric discharge capable of provoking the initial fusion of both of them as well as the resulting disintegration and release of energy and radiations, causing a chain reaction that would continue for long time, to the tune of trillions of years. Fig. 29 illustrates this sequence: 1) two capacitors, deuterium (on the left) and tritium; 2) the fusion of these two atoms (or capacitors); 3) the disintegration, in the photo represented by only an alpha (α) and beta (β) particles, the latter by two electrons. Thus, the mass of matter that contains the atoms of deuterium and tritium–helium-5 (He2/5) isotope and the particles emitted through disintegration–embody the following possibilities of reacting to each other: a) an alpha particle upon capturing two electrons (beta particles), is completed and creates the first stable deuterotome, that is, helium-4 (He2/4); b) the deuterotome of helium-4 (He2/4) upon capturing an alpha particle gives rise to berilium-8 (Be4/8) deuterotome; the synthesis continually leads to the formation of deuterotomes by capturing alpha particle with deuterotomes number 1 increasing two by two and the mass number four by four, forming, then, a phylogenetic lineage of calcium-40; c) helium-5 (He2/5) deuterotome itself can have a few units maintained whole, despite its unstable nature, and capable of capturing neutrons (n) emitted as a result of the disintegration of another helium-5 deuterotome and from helium-6 (He2/6) isotope. Because this deuterotome has unstable nature, it emits a beta particle (β) and changes into lithium-6 (Li3/6) deuterotome. Lithium-

6 is a stable deuterotome, but because it exists in the environment at a critical temperature and pressure and in the presence of high concentrations of neutrons, it captures a unit and synthesizes the isotope with a mass number equal to 7 (Li3/7).

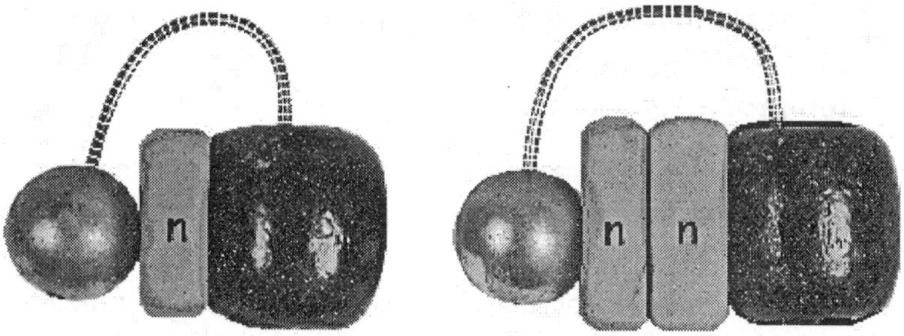

Fig. 29.1 – Illustration of deuterium (on the left) and tritium atoms. The deuterium atom has one neutron between the proton and the electron. Tritium has two neutrons between them.

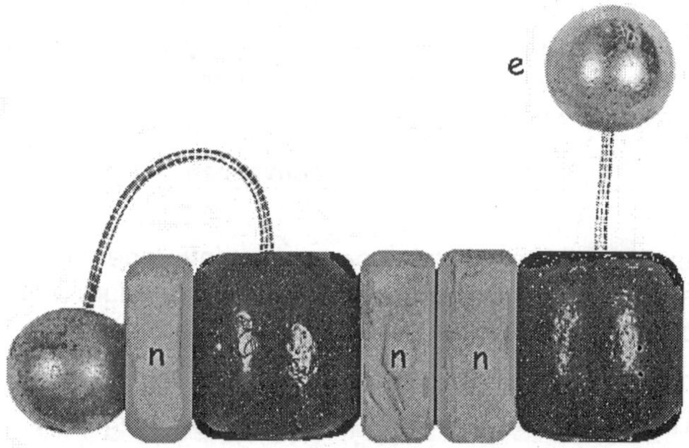

Fig. 29.2 – Illustration of the nuclear fusion between deuterium and tritium atoms, forming unstable helium-5 (He2/5) isotope, which will disintegrate into alpha and beta particles, neutron and gamma rays photons. During the disintegration, helium-3 (He2/3) isotope will also be formed. A few He2/5 units will remain intact and capable of capturing neutrons.

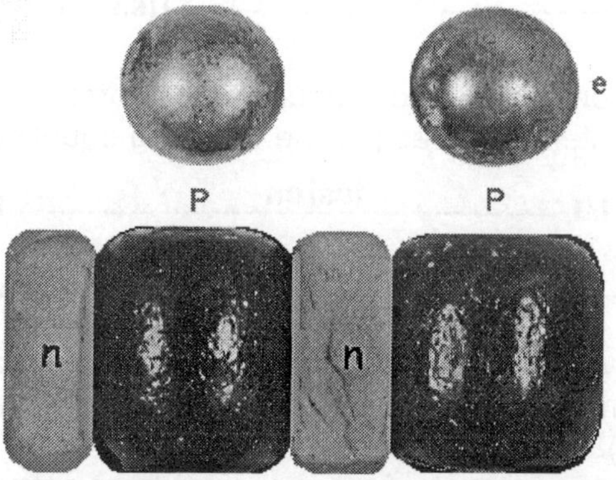

Fig. 29.3 – Illustration of alpha (α) and beta particles (β) resulting from the disintegration of helium-5 isotope (He2/5). Two electrons, both placed on top of the photo

represent the beta particles. Alpha particle is formed by two protons (P) and two neutrons (n). The neutrons function as dielectric or insulator.

3.5 – Mechanisms of synthesis of deuterotomes

3.5.1 – Proton-proton cycles

Whenever a stellar unit begins a stable period, a great level of contraction is upheld by internal pressure, due to the radiations (thermal photons and gamma rays photons) traveling from its interior towards the surface.

This stellar unit continues to produce energy throughout the reactions of nuclear fusion at a rate determined by its mass. For a star, such as the Sun, the stable period can last ten billion years (10^{10} years). For a star 10 to 100 times heavier than the Sun, this stable period of burning hydrogen can last less, somewhere around 100 million years.

The basic reaction of nuclear fusion, resulting from the fusion of two atoms of light hydrogen (H1/1), originating a transitory short-lived bi-hydrogen (H2/2), followed by the beta decay process, with the emission of positive beta particle (β+) or positron and neutrino (v), can be symbolically shown as follows:

$$\text{H1/1} + \text{H1/1} \xrightarrow{\text{fusion}} \text{H2/2} \xrightarrow{\text{conversion}} \text{H1/2} + e^{+} + v$$

This process involves the conversion of a proton into a neutron and is similar to the beta decay process. As soon as the heavy hydrogen or deuterium atom (D1/2) is formed, the immediate reaction that can occur as follows:

$$\text{H1/2} + \text{H1/1} \xrightarrow{\text{fusion}} \text{He2/3} + \gamma$$

Helium-3 (He2/3) isotope is formed during this reaction with two protons and one neutron, which go through a fusion process, as shown in the below reaction:

$$\text{He2/3} + \text{He2/3} \longrightarrow \text{Be4/6}$$

$$\text{Be4/6} \longrightarrow \text{He2/4} + 2\text{H1/1}$$

The chemical reaction shown below requires four atoms of light hydrogen to produce one helium-4 (He2/4) deuterotome, as per the following equations:

$$4\text{H1/1} \xrightarrow{\text{fusion}} \text{H4/4}$$

$$\text{H4/4} \xrightarrow{\text{conversion}} \text{He2/4} + 2e^{+} + 2v + 2\gamma$$

In the reaction above there is production of two positrons (anti-electrons) by a beta decay process, by the emission of beta positive particles (β+), two neutrinos (v) and two gamma rays photons (γ).

In order for the Q value (excess of mass energy) to be calculated pursuant to atomic masses, four electrons must be added to the first term of the equation (on the left) to form four neutral hydrogen atoms. To balance the reaction four electrons must be added to the second term (on the right); two of those electrons will blend with the deuterotome of helium-4 (He2/4) and the other two will combine with the two positrons, in keeping with the equation below:

$$2\,e+ \quad + 2\,e^{-} \longrightarrow 2\gamma$$

These two additional gamma rays photons are energy resulting from this reaction. Because the two positrons (e+) disappear during this process, the only mass that remains is formed by four atoms of light hydrogen (H1/1) plus one helium-4 (He2/4) deuterotome. The **Q value**
equation:

$$Q = (im - fm) \times c^2$$

Q is the excess of mass energy,

im is the initial mass of light hydrogen atom,

fm is the final mass of helium-4 deuterotome,

c² is the square of light speed (expressed in 931.5 MeV /μ).

Thus, Q = (4 x 1.007825μ – 4.002603μ) x 931.5 MeV/μ). As per the calculation, the Q value will be equal to 26.7 MeV.

Each fusion reaction, releases approximately 26.7 MeV of energy to form a helium-4 (He2/4) deuterotome.

proton-proton cycles, most likely account for the source of solar energy. The said types of reactions, however, seem not to be the primary source of energy from fusion in many stellar units because the former reaction–positive beta decay (β+) type–only occurs at a very long time span, which make this process unlikely to happen on a large scale. However, these reactions do provide the amount of irradiated energy by the stellar units, releasing 26.7 MeV (mega electron-volts) of energy when four atoms of light hydrogen are blended to form the deuterotome of helium-4. Stellar units similar to the solar ones produce most of their energy by means of the proton-proton cycles. Nevertheless, such units do not have enough heat to do the **carbon cycle** or, most important, carbon (C6) is not available in their masses.

Depending on the availability alpha particles (He2/4), three different proton-proton cycles can occur and be sources of alpha particles and mechanism of deuterotomes synthesis. The basic chemical reactions for the three types of cycles are as shown below:

$$H1/1 + H1/1 \xrightarrow{\text{fusion}} H2/2 \xrightarrow{\text{conversion}} H1/2 + e^{-} + \nu$$

$$H1/1 + H1/2 \xrightarrow{\text{fusion}} He2/3 + \gamma$$

The three types of proton-proton cycles are provided with the same reactions, that is, until the helium-3 (He2/3) isotope is formed. Starting from it, variations in the pathways for nuclear fusion to occur based on the availability of the atoms that participate in the fusion process. The three likely fusion pathways are as follows:

a) He2/3 + He2/3 ------------------ > Be4/6
Be4/6 ----------------------------- > He2/4 + 2H1/1

b) He2/3 + He2/4 ------------------ > Be4/7 + γ
Be4/7 + e ------------------------- > Li3/7 + v
Li3/7 + H1/1 --------------------- > 2He2/4

c) He2/3 + He2/4 ------------------ > Be4/7 + γ
Be4/7 + H1/1 ------------------- > B5/8 + γ
B5/8 ------------------------------ > Be4/8 + e+ + v
Be4/8 ---------------------------- > 2He2/4

Any of the three proton-proton cycles has the same basic reactions up to the synthesis of He2/3 deuterotome composed of two protons and one neutron which also emits one positron (e+), one neutrino (í) and releases gamma radiation (γ).

In the first fusion pathway (letter **a**), two helium-3 deuterotomes go into fusion and give rise to berilium-6 (Be4/6) deuterotome, which is decomposed into one alpha particle (He2/4) and two atoms of light hydrogen (H1/1).

In the second fusion pathway (letter **b),** the fusion (through collision) between the helium-3 isotope and helium-4 determines the formation of berilium-7 deuterotome (Be4/7); upon an electron capture (e^-) the latter has one proton transformed into neutron, thus reducing the deuterotomic number from 4 to 3 without altering the mass number and consecutive emission of neutrino (v). In the final stage, lithium-7 (Li3/7) fuses with an atom of light hydrogen (H1/1) resulting in berilium-8 (Be4/8) deuterotome, which disintegrates into two alpha particles (He2/4).

In the third fusion pathway (reaction **c**), the initial stages of the fusion process up to Be4/7 are the same, but here berilium-7 fuses with one atom of light hydrogen (H1/1), giving rise to boron-8 deuterotome (B5/8) and the emission of gamma rays photons. This deuterotome undergoes positive beta decay (β+) upon the emission of one positron or anti-electron (e+) and one neutrino (v). Berilium-8 deuterotome disintegrates into two alpha particles (He2/4).

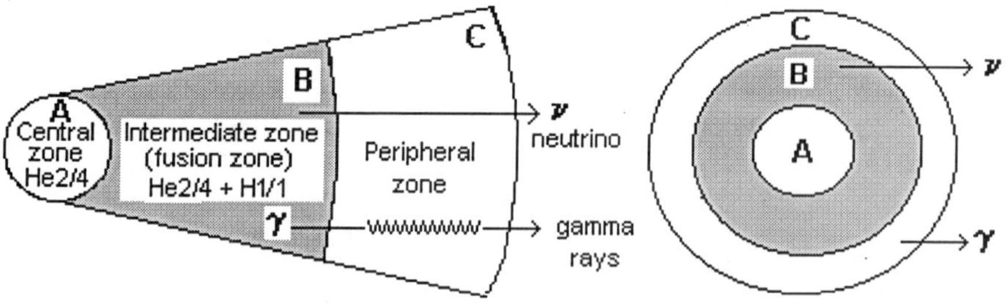

In all likelihood, most of the solar energy that is produced results from reactions of proton-proton cycles, described in **b** reaction. Noble gas helium, present in the cloud of initial gas concurrently with that produced through fusion, forms a nucleus A in the center of the stellar unit, since helium (He2/4) is heavier than light hydrogen (H1/1); light hydrogen lies on zone C; between these two layers is an intermediate zone B where helium and hydrogen come into contact and where reactions of nuclear fusion also occur. The foregoing scheme shows the three zones in the stellar unit.

The fusion energy is released mainly in the form of gamma ray photons (95%) and neutrino (5%), reaching the stellar surface and irradiated throughout space. Gamma rays, submitted to Compton dispersion inside the solar unit due to its dense material, may take 10^5 to 10^6 years to reach the surface. Inside a star, gamma rays lose energy until they turn into photons (quanta) of visible light. Therefore, what becomes exteriorized in the Sun is a characteristic than occur on its surface, not in its interior. Neutrinos, for their turn, because they have a small cross-section for interaction with matter, roam directly across the solar matter at the speed of light. Accordingly, the neutrino that arrives at the earth atmosphere stems from reactions of nuclear fusion inside the Sun. The neutrino captured by deuterotomes on a small scale, causes nuclear alteration and consequent beta decay as per the equation below:

$$v + Cl17/37 \longrightarrow Ar18/37 - e^-$$

This reaction shows the transformation of clorine-37 deuterotome (Cl17/37) into argon-37 (Ar18/37) deuterotome and the emission of a negative beta particle (β^-).

3.5.2 – Carbon cycle

The most probable sequence of reactions in the carbon cycle is the one where this element functions as a catalyst in the form of C6/12, since no carbon is consumed during the cycle. This cycle also forms helium-4 (He2/4) deuterotomes, consuming four protons, two of them being used in the synthesis of helium deuterotome and the two other are transformed in positrons (anti-electrons). The sequence of the transformations is schematized below.

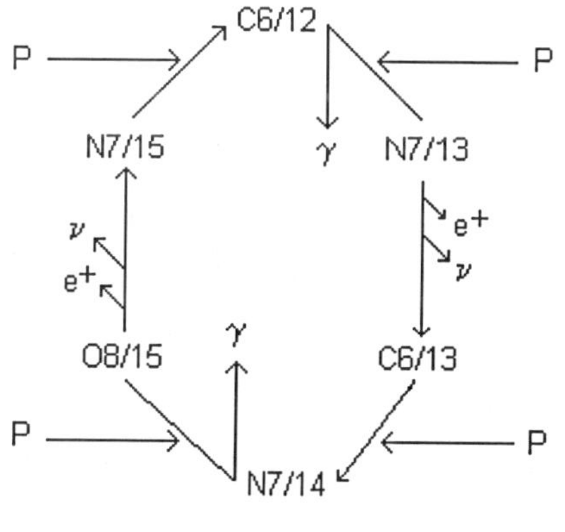

Fig. 30 – Sequence of transformations in the carbon cycle.

Carbon-12 (C6/12) has a catalyst function, because this deuterotome is restored by the end of carbon cycle; there is neither production nor consumption of C6/12 in these reactions, but its presence enables this sequence of reactions to take place at a rate greater than that of the proton-proton cycles. The carbon cycle process can be shown by the following equation:

$$4H1/1 \xrightarrow{\Delta} He2/4 + 2e+ + 2\nu + 2\gamma$$

Coulomb repulsion force existing among the atoms of hydrogen (H1) and carbon (C6) is greater than that between two nuclei of hydrogen; consequently, the fusion of two helium (He2/4) deuterotomes requires far more energy than that of two atoms of hydrogen. A greater thermal energy and a corresponding higher temperature are necessary in the carbon cycle. The carbon cycle is probably more important at the temperature of $20,10^6$ K, while the temperature in the interior of the Sun is lower, around $10,10^6$ K (Kelvin). When all the stellar hydrogen is converted into helium, the Sun will contract and will heat up until the temperature reaches a level high enough to enable helium to burn via the fusion processes in keeping with the chemical reactions below that leads to carbon-12 (C6/12) synthesis.

3He2/4——————>C6/12

When all helium is depleted, a higher temperature will enable the fusion of carbon to form heavier elements, similar to magnesium-24 (Mg12/24). This process will go on until the synthesis of iron-56 (Fe26/56) deuterotome can be reached; beyond this point, no energy is gained though the nuclear fusion process.

3.5.3 – Synthesis by "neutron capture"

Elements with a deuterotomic number up to 26 (Fe26) were synthesized by means of fusion reactions in stellar units. However, beyond iron (Fe26), the synthesis via fusion reactions is not energetically favorable. The elements above it were synthesized by neutron capture, producing isotopes with the excess of one neutron according to the chemical equations below:

$$Fe26/56 \xrightarrow{+n} Fe26/57 \text{ (stable)}$$

$$Fe26/57 \xrightarrow{+n} Fe26/58 \text{ (stable)}$$

The synthesis of elements above iron (Fe26) goes on by the mechanism type "neutron capture and emission of beta particle" up to uranium-238 (U92/238). The sequence below shows the stages in this type of synthesis:

$$Fe26/58 + n \longrightarrow Fe26/59 \xrightarrow{-\beta} Co27/59 \text{ (stabe)}$$

$$Co27/59 + n \longrightarrow Co27/60 \xrightarrow{-\beta} Ni28/60 \text{ (stable)}$$

All of the transuranium (U92) elements are produced artificially through synthesis reactions induced by emissions of atomic projectiles, such as slow neutrons, and via the beta decay process. All the transuranium elements are radioactive and do not constitute part of the earth natural matter; many of them have very short half-life of some minutes or seconds, and the amount produced does not go beyond a few deuterotomes. However, most of these elements up to 107 have not been produced in quantity enough to enable studies in depth as to their physical and chemical properties.

The great instability in transuranium elements results from an increase in Coulomb repulsion forces of nuclear protons. These elements undergo a radioactive degradation process via alpha decay type or through spontaneous fission (without first there having been a neutron capture). However, there are remarkable theoretical evidences which point out–for reasons based on the nuclear structure–that elements with deuterotomic number around 114 and neutrons number around 184 would be stable against alpha decay, beta decay and spontaneous nuclear fission. Due to the tiny amounts of elements with deuterotomic numbers ranging between 100 and 114, it is unlikely that the neutron capture and the emission of beta particles (β) would be a process used to produce element 114 to be observed. Another nuclear reaction using a projectile-element of intermediate mass number or small value such as argon-40 (Ar18/40) to bombard a heavy deuterotome like curium-248 (Cm96/248), for example, is not viable, because there are not enough neutrons available to supply the 184 neutrons necessary for the deuterotomic stability. It is necessary to use nuclei rich in neutrons as well as atomic projectiles as deuterotomes such as U92/238 as projectile or even U92/238 as target, in the hope that an intermediate stage–a composed nucleus–might be formed which would stabilize the isotope with a deuterotomic number equal to 114, possibly through nuclear fission.

A simple mechanism, "neutron capture" type, which creates isotopes, is present in a few reactions with an immediate stop in the synthesis progression. The following are typical examples:

Ti22/46	+n ———	> Ti22/47	Hg80/200	+n ——	> Hg80/201
Zn30/66	+n ———	> Zn30/67	Pb82/206	+n ——	> Pb82/207
Se34/76	+n ———	> Se34/77	Th90/230	+n ——	> Th90/231
Os76/188	+n ———	> Os76/189	U92/234	+n ——	> U92/235

3.5.4 – Synthesis by "neutron capture and beta particle emission"

A pattern of deuterotomic synthesis is based on the property of uranium-238 (U92/238) deuterotome of capturing slow neutrons, increasing its mass number, being changed into uranium-239 (U92/239) and emitting a beta particle (β), thus raising the deuterotomic number and being changed into neptunium-239 (Np93/239), an isobar of uranium-239. This can be schematically represented as follows:

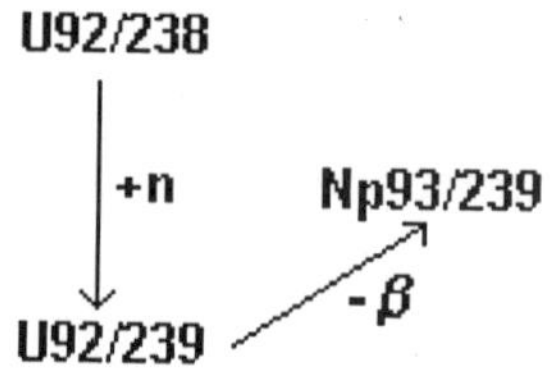

This is a standard mechanism that fits the process of deuterotomes synthesis from the isotope of helium-5 (He2/5) which via capture of a neutron has changed into an isotope with a mass number equal to 6 (He2/6) and by emitting a beta particle has changed into lithium-6 (Li3/6). As the transformation from He2/5 into Li3/6 was total, this helium isotope became extinct, not longer found in nature. Another example can be given about this type of reaction: sulfur-36 (S16/36) by capturing one neutron changes into sulfur-37 (S16/37); the latter emitted a beta particle (β) and changed into clorine-37 (Cl17/37). Likewise, isotope S16/37 changed totally into Cl17/37 and became extinct, no longer found in nature either.

A deuterotome, upon capturing one neutron, creates an isotope with a greater mass number; the presence of neutron (or one dielectric) increases the **d** value in the capacitance equation, reducing the capacitance of the deuteron unit that receives the neutron; and consequently, the total capacitance of the deuterotome as well, whichever the binding form among the plates–series-equivalent or parallel-equivalent. The reaction below is a typical example.

Bi83/209 + n ----------------------------> Bi83/210 ---------- -β ---------- -γ ----------> Po84/210

Bismuth-209 (Bi83/209) deuterotome captures one neutron (n), unbalances the capacitance of both the receiving deuterian unit; it causes a nuclear excited state and then becomes an isotope with greater mass number (Bi83/210)–unstable and radioactive– due to the entrance of the neutron. This isotope emits one beta particle (β) from the electron integrating the neutron; the latter, devoid of electrons acquires a proton state and increases the Z deuterotomic number by 1; the nucleus in excited state emits nuclear gamma rays.

This type of synthesis mechanism is present in at least 64 deuterotomes formations, as can be seen in the “Picture of possible pathways for deuterotomes synthesis”. It is a mechanism of dual function: it increases the mass number and raises the deuterotomic number, both by 1.

However, the isotope created by the capture of a neutron is totally transformed into a deuterotome of greater deuterotomic number, albeit an isobar. When the overall transformation of a neutron receiving deuterotome occurs, it becomes an “extinct isotope”. This phenomenon, however, did not happen to all of the isotopes that had undergone that synthesis stage; a few still exist and became something like “residual isotope”, still radioactive. Cadmium-113 (Cd48/113), for example, which originates from cadmium-112 (Cd48/112) that had captured a neutron, did not entirely change into indium-113 (In49/113), a part of 4.5% of the total element iridium is encountered in nature. Cd48/113 isotope accounts for 12.3% of the total of cadmium deuterotomes. It is, thus, a “residual isotope" from what should be an "extinct isotope." This is an example that can be seen in the "Picture of possible pathways for deuterotomes synthesis".

A shift from the synthesis pathways happened to calcium-46 (Ca20/46), present at a rate of 0.0033% of the total calcium encountered in nature. The said isotope had captured one neutron and created isotope Ca20/47. According to the natural order of synthesis, it should have emitted one beta particle and changed into scandium-47 (Sc21/47). But this did not happen. Rather another neutron was captured, creating calcium-48 (Ca20/48) isotope, which represents 0.185% of the total.

Another shift from the synthesis pattern happened to lutetium-175 (Lu71/175) deuterotome which corresponds to 97.5% of the total, whose radioactive transformation should have ceased at this stage because it originated from the mechanism "neutron capture, emission of beta particle". However, this deuterotome captured one neutron and created its own isotope one–lutetium-176 (Lu71/176)–that accounts for 2.5% of all deuterotomes encountered in nature.

By way of this synthesis mechanism, four deuterotomes should be abundant and stable, but they are regarded as radioactive isotopes with a low concentration and only references are made as to their half-lives. These isotopes are: Nb41/95, Te52/129, Cs55/135 and La57/137. Their position can be verified in the "Picture of possible pathways of deuterotomes synthesis", as follows:

1) - Zr40/93 (3 My or million of years) formed Nb41/93 (100%) and Zr40/95 (64 days) should have formed Nb41/95 (35 hours),

2) - Te52/127 (9.4 hours) formed I53/127 (100%) and Te52/129 (69 months) should have formed I53/129 (16 My),

3) - Xe54/133 (5.25 days) formed Cs55/133 (100%) and Xe54/135 (9.1 hours) should have formed Cs55/135 (3 My),

4) - Ba56/139 (71.7%) formed La57/139 (99.911%) and Ba56/137 (11.2%) should have formed La57/137 (0.06 My).

Therefore, even through this mechanism is available and instrumental in the synthesis of heavier deuterotomes, not always is it present systematically. The reason is not self-evident, but the fact that a deuterotome can capture a neutron requires that both be available in appropriate concentration and in the same environment where the synthesis must take place; the inadequate concentration of one or the other may be have been one of the reasons. Consequently, the radioactive phenomenon of beta decay will take place where the radioactive deuterotome is lying because this phenomenon in entirely dependent on the electron, which is neutralizing the unit that makes up the neutron, being, as a result, an intradeuterotomic phenomenon not dependent on the concentration or the surrounding environment.

3.5.5 – Synthesis by "capture of alpha particles"

This is a simple mechanism for the synthesis of deuterotomes and at the same time appears to have been a quick synthesis mechanism, for upon each capture of an alpha particle (He2/4); the deuterotome then formed starts to have two more protons and two neutrons than the deuterotome that originated it. Among lighter deuterotomes, the synthesis process ceases after a string of nine or 10 capture undergoing a stop, the reason of which fails to be self-explanatory. The only series that was composed of a synthesis sequence exclusively based on the capture of alpha particles was the one ranging from helium-4 (He2/4) up to calcium-40 (Ca20/40). All the other series are associated with neutrons being captured in several deuterotomes in each series. In a sequence, the maximum number of deuterotomes that have captured neutron did not exceed 10 which is the linage ranging from oxigen-18 (O8/18) up to nickel-58 (Ni28/58); see "Table of possible pathways for deuterotomes synthesis".

The "Table of synthesis pathways" underlines the existence of empty gaps between two deuterotomes with even deuterotomic numbers–e.g. Zn30/68 and Ge32/72)–which indicates that the synthesis process in these phylogenetic lineages happened via capture of alpha particles. The deuterotome generated by this mechanism might be stable or not. Depending on the various factors that can be instrumental in the stability of the deuterotome then formed, one of them may be the capacitance rate of its deuterium units and the total capacitance of the deuterotome. A nuclear reaction controlled by reactors used in the process of obtaining astatine-211 isotope (At85/211) can be symbolically represented as per the following equation: Bi83/209 + He2/4—> At85/213 – 2 n——> At85/211.

Bismuth-209 (Bi83/209) isotope captures an alpha particle (He2/4), raises the total capacitance rate of the deuterotome then formed which becomes a transitory, non-isolating, isotope of astatine-213 (At85/213) that, successively, emits two neutrons, resulting in the radioactive isotope of astatine-211 (At85/211) with a short half-life of 7.21 hours. The capacitance rate of the units that ejected neutrons increased, because the distance **d** among the plates (protons) decreased. The total capacitance rate also increased, because the dielectric thickness shortened, and also because of the presence of two more receiving plates–the two protons of the alpha particle previously captured.

From the same element–e.g. iron-58 (Fe26/58)–the performance of the two types of mechanism ("alpha capture" and "neutron capture and beta particle emission") results in the synthesis of different deuterotomes as regards their mass number. The scheme below shows these two types of mechanisms.

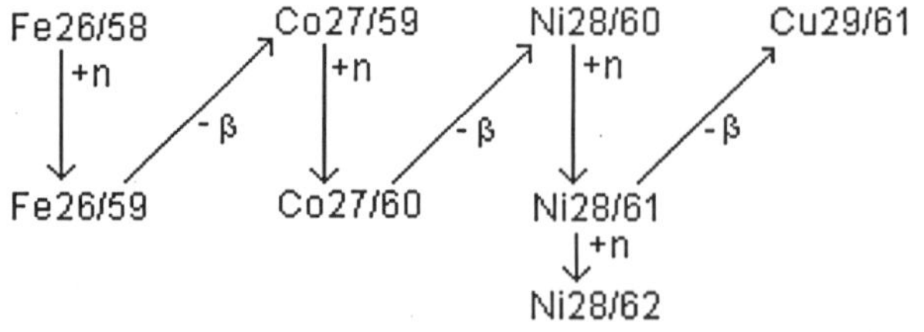

The two synthesis mechanisms point out that the capture of alpha particles gives rise to nickel isotope with a mass number greater than the capture of two neutrons and two emissions of beta particles. The results arising from these mechanisms do not overlap; each one forms deuterotomes that uphold one another, in relation to both the deuterotomic number and the mass number. As a result, isotopes and isobars are created with a large number of capacitance values, whether deuterons or deuterotomes. This diversity raises the likelihood of chemical reactions to occur, and produce multiple variations as regards the physical properties of these deuterotomes. Elements that participate in the composition of biological chemical substances, in all of their levels of complexity, from a simple molecule of urea to the complex molecule of a deoxyribonucleic acid, for example, can offer subtle specificities by the capacitance values of their genetic poles, the number of free valences in their chemical poles, the electromotive force of their valences, the capacitance of their deuterons and deuterotomes, in creating and maintaining biodiversity.

The radioactivity of elements that emit beta particles (β) is spontaneous and results from the instability brought about by the neutron's entering in the deuterotome. This radioactivity is not dependent on the existence of the "electronic furnace" or any other external factor. The emission of beta particle is a radioactive physical phenomenon whose origin is intradomicile. This characteristic enables the element to emit beta particles wherever the deuterotome lies, for radioactivity is a property not adversely affected by external factors–e.g. variations in the temperature and pressure, electric field, magnetic field, etc. On the other hand, the capture of alpha particles or neutrons is entirely dependent on external factors; in other words, it is dependent on how close together the deuterotomes and the particles are. If there had not been a large number of basic elements–capturers of alpha particles–they would be far in excess since they are continuously formed throughout the process of nuclear fusion.

3.5.6 – Synthesis by "neutron capture, beta particle emission and new neutron capture"

This type of mechanism is present in many deuterotomes and a few examples can be given, where lithium (Li3), boron (B5), nitrogen (N7), potassium (K19), lutetium (Lu71), lanthanum (La57) and tantalum (Ta73), besides others, are present:

1–He2/5 + n -------- > He2/6 – β --------------> Li3/6 + n ----------- > Li3/7 (92.5%)
2–Be4/9 + n -------- > Be4/10 – β ------------> B5/10 + n --------- > B5/11 (80%)
3–C6/13 + n -------- > C6/14 – β -------------> N7/14 + n --------- > N7/15 (0.38%)
4–Ar18/38 + n ----- > Ar18/39 – β ----------> K19/39 + n -------- > K19/40 (0.01%)
5–Yb70/174 + n --- > Yb70/175 – β --------> Lu71/175 + n ----- > Lu71/176 (2.5%)
6–Ba56/136 + n --- > Ba56/137 – β ---------> La57/137 + n ----- > La57/138 0.089%)
7–Hf72/178 + n --- > Hf72/179 – β ---------> Ta73/179 + n ----- > Ta73/180 (0.0123%)

Isotopes He2/6, Be4/10, C6/14, Ar18/39 and Yb70/175, referred to in the examples above, are considered "extinct isotopes" because they changed almost totally into the deuterotomes that resulted from the emission of beta particles from each one. When present they are radioactive such as beryl (Be4) and carbon (C6), for example: helium-5 and helium-6 are not available in nature.

3.5.7 – Ratio between neutrons and protons in deuterotomes

As far as the structured view is concerned, one or two neutrons lying among the plates can form the atomic capacitor. In deuterium there is one neutron, in tritium there are two. From helium-4 (He2/4) onwards, the ratio is one neutron per one proton (1n: 1p), until heavy deuterotomes are reached, where this ratio is wider; for uranium-238 (U92/238) the ratio is 1.58n: 1p. An increase in the number of neutrons reduces the total capacitance of the deuterotome, as can be seen in the chart of radioactive degradation of both radium and actinium families.

An atomic capacitor can be composed of a number of neutrons that reach a ratio of 10 neutrons: 2 protons, that is, 5n:1p which is present in the "alpha-appendix" of uranium-238, in the radioactive family of radium.

3.5.8 – Synthesis of some transuranium deuterotomes

From uranium (U92) onwards, all of the deuterotomes known to date have been artificially synthesized and are all radioactive, complying with the very structure of the uranium deuterotome, that is, they have a genetic pole constituted by noble gas xenon (Xe54), an "electronic furnace" formed by 32 electrons that determines both the radioactive property and the chemical pole where chemical reactions are processed. The elements known to date are the following: neptunium (Np93/237), plutonium (Pu94/242), americium (Am95/243), curium (Cm96/247), berkelium (Bk97/247), californium (Cf98/251), einsteinium (Es99/254), fermium (Fm100/253), mendelevium (Md101/256), nobelium (No102/253), lawrencium (Lr103/257), dubnium (Dh104/261), juliotium (Ji105/262), rutherfordium (Rf106/263), bohrium (Bh107/264), hahnium (Hn108/265) and meitnerium (Mt109/266).

The mechanism type "neutron capture, beta particle emission" processes the synthesis of transuranium deuterotomes from uranium-238. The chart below shows the initial sequence of the synthesis.

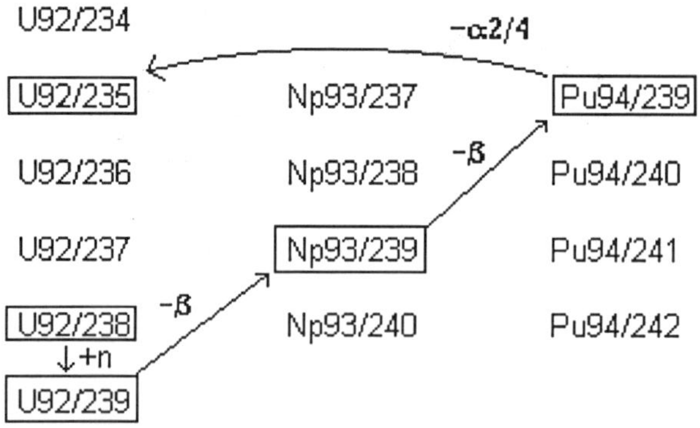

A slow neutron (**n**) used as an atomic projectile is captured by uranium (U92/238), resulting in an isotope with a mass number equal to 239 (U92/239); the latter emits a beta particle (β) and changes into neptunium-239 (Np93/239), an isobar of uranium-239. The neptunium then formed also emits a beta particle (β), generating plutonium-239 (Pu94/

239), also an isobar of that deuterotome. The emission of one alpha particle (α 2/4) by plutonium-239 produces uranium-235 (U92/235). Thus, throughout the process of artificial synthesis of transuranium deuterotomes, the natural radioactive degradation is present via emission of alpha particles. This radioactive property results from the property of the "electronic furnace", which remains unaffected in that deuterotome and in other transuranium deuterotomes as well.

3.5.9 – Induced synthesis of some deuterotomes

The initial synthesis of matter is processed from the lentons that overlap one other, creating lentonian plates. The electron is a unit formed by **n** x 1 plate; the proton is another unit formed by **n** x 1,836 plates. From these two electric units originate the atoms of hydrogen and the neutron. From the latter, stem both the deuterium and tritium atoms via capture, these two atoms make up atomic capacitors. The fusion of these two atoms or capacitors (nuclear fusion) creates the isotope of helium-5 (He2/5) deuterotome, unstable, which is decomposed into atomic particles (alpha, beta and neutrons) and emitting gamma rays, which will carry out the synthesis of deuterotomes through several possible pathways.

From deuterotomes already available, new elements can be synthesized by artificial means–called transmutation–whereby deuterotomes with both higher deuterotomic number and greater mass number can be obtained, however, in no way does that mean that matter can be created since there is a balance of energy and mass on the two terms of the transmutation equations. Nearly all particles (alpha, neutron, proton) are present in the mechanism of artificial synthesis in addition to atoms of heavy hydrogen (D1/2) and gamma radiations (γ). In the illustration below in parentheses, the first note refers to the atomic projectile used and the second to the particle emitted.

1 – (α, n) Be4/9 +α 2/4 ----------------------> C6/12 + n (Chadwick bombardment)
2 – (α, n) Al13/27 +α 2/4 ---------------------> P15/30 + n
3 – (n, β) U92/238 + n-----------------------> U92/239 + β
4 – (n, π) S16/32 + n -------------------------> P15/30 + π + 2n
5 – (π, α) F9/19 +π ---------------------------> O8/16 + α 2/4
6 – (π, n) Ar18/40 + π -----------------------> K19/40 + n
7 – (π, γ) Be4/9 + π---------------------------> B5/10 + γ
8 – (D, γ) C6/12 + D1/2 ----------------------> N7/14 + γ
9 – (D, n) B5/11 + D1/2 ----------------------> C6/12 + n
10 – (γ, n) D1/2 + γ ----------------------------> H1/1 + n (photonuclear effect)
11 – (n, γ) H1/1 + n ----------------------------> D1/2 + γ

During the bombardment of a deuterotome by an accelerated deuterium (D1/2), the neutron is the particle that works, which breaks loose from the deuterium through shock. The neutron penetrates the deuterotome but the proton does not participate in that reaction. From the viewpoint of transmutation, the bundle of deuterons acts as a bundle of neutrons. Gamma rays also cause transmutations–they emit photons, which are absorbed by the deuterotome whose nucleus is in excited state. In example 10 above, the deuterium is bombarded by gamma rays that are decomposed into one atom of light hydrogen (H1/1) and one neutron. This phenomenon is called photonuclear effect. However, the opposite also happens: if the light hydrogen is bombarded by neutrons, deuterium will then be formed and there will be emission of gamma rays as per equation 11.

Transmutations caused by neutrons, acting as atomic projectiles, have a higher rating the slower the speed of these neutrons. The speed of neutrons will be slower if they cross thicknesses of moderator's substances such as paraffin, heavy water (D_20) etc. The moderators of atomic reactors have the property of slowing down the speed of fast neutrons turning them into slow ones.

CHAPTER 4

Phylogenetic lineage of deuterotomes and Table of possible ways for the synthesis of deuterotomes

Deuterotomes consist of juxtaposed deuterons so as to form a sequence. Every noble gas deuterotome precedes a certain string of deuterons. The extremity where a certain noble gas lies is called **genetic pole**. Given that deuterotomes are synthesized in accordance with a line or series and are dependent on the properties of the respective genetic poles, one can say that each is provided with a phylogenetic lineage (from the Greek **phylon**, race, people, and **genesis**, origin). Given that each noble gas has a specific capacitance value, capable thus of guiding the synthesis of deuterotomes that follow ahead; it can be inferred that the capacitance will be the source of electric load in each deuterium that will make up the deuterotome; it can be also said that the synthesis of matter has complied or complies with an atomic genetic code.

The process of matter synthesis must have occurred in accordance with ground rules that helped maintain its flow based on the generator of neutrons and particles (alpha and beta) resulting from the reaction of a nuclear fusion between deuterium and tritium, producing unstable helium-5 isotope broken into atomic particles. Consequently, with the aid of a source of neutrons and alpha particles, the initial basic matter (He2/3, He2/4, He2/5 and He2/6) was then able to form deuterotomes within a well-defined phylogenetic lineage by way of the synthesis mechanisms–alpha particle capture, neutron capture and emission of beta particles. The following lineages can be underlined:

1) – Natural isotope whose existence cannot be denied. For instance, clorine-34 (Cl17/34), potassium-37 (K19/37) etc.,

2) – A great number of isotopes existed temporarily and became an isobar deuterotome of another with a higher deuterotomic number (see a few examples of "extinct isotope and/or residual isotope"),

3) – Isotopes that are not largely found in nature or are radioactive, even though they take part in the phylogenetic lineage of deuterotomes. These isotopes are estrontium-90 (Sr38/90), zirconium-88 (Zr40/88), cerium-139 (Ce58/139), cerium-144 (Ce58/144) and neodymium-140 (Nd60/140).

4.1 – Ground rules in the synthesis processes

One can identify some ground rules encountered throughout the synthesis processes, which help understand the paths most likely used and the progression interruption (stop) that happens to a given lineage. Evidence that an interruption (stop) in the progression is really true lies in the fact that the isotope of uranium-184 (U92/184) deuterotome, for example, does not exist. It would be the natural outcome from the synthesis due to the capture of alpha particles by helium-4 (He2/4) with no stop. These ground rules are as follows:

1) Heavier isotopes of a deuterotome have a far greater capability of capturing neutrons,

2) When a neutron is captured by an isotope from a deuterotome with both odd number and mass, originates an isotope with a greater mass is to ensue bound to be entirely changed into a deuterotome with a higher deuterotomic number, falling thus into the category of "extinct isotope" or a "residual isotope", should it still remain radioactive in minimal quantity. The following are examples of these isotopes: germanium-75 (Ge32/75), samarium-153 (Sm62/153) etc.,

3) Deuterotomes with even number and mass, capable of capturing alpha particles, can easily capture 5 to 7 particles when a synthesis stop occurs. In the example below, the capture of six alpha particles from the genetic pole formed by argon-40 (Ar18/40), leads the synthesis as far as zinc-64 (Zn30/64) where the synthesis stops,

$$Ar18/40 \xrightarrow{+\alpha} Ca20/44 \xrightarrow{+\alpha} Ti22/48 \xrightarrow{+\alpha} Cr24/52 \xrightarrow{+\alpha}$$

$$\longrightarrow Fe26/56 \xrightarrow{+\alpha} Ni28/60 \xrightarrow{+\alpha} Zn30/64 \text{ (stop)}$$

4) The isotope of heavier deuterotomes, with even number and mass, increases its mass number when it captures a neutron, changes into an odd number and becomes unstable; after emitting a beta particle (β), it then becomes a deuterotome with a greater deuterotomic number, increased by 1 (one),

5) Deuterotomes with odd mass number are more often encountered at the end of a series or phylogenetic lineage like terbium-159, for example. The following example shows this ground rule:

Ga64/158 + n———> Ga64/159 – β ———> Tb65/159 (stop),

6) Deuterotomes that capture neutrons and emit beta particles give rise to deuterotomes with both odd number and mass, lying in the extremity of a phylogenetic lineage that undergoes an interruption in the synthesis progression (stop). A few examples: fluorine-19 (F9/19), sodium-23 (Na11/23), aluminum-27 (Al13/27), gold-197 (Au79/197), bismuth-209 (Bi83/209) etc.,

7) The series of deuterotomes synthesized due only to the capture of alpha particles is short. The model that exemplifies the said series is the sequence ranging from lithium-6 (Li3/6) to calcium-40 (Ca20/40). The synthesis progression is interrupted in the latter (stop),

8) The constant and regular empty interval between deuterotomes with pairs deuterotomic number and deuterotomic mass is proof enough that the synthesis of

deuterotomes in that phylogenetic lineage happened through the mechanism alpha particles capture; in the sequence O8/16 —> Ne10/20 —> Mg12/24 —> Si14/28, the intermediate deuterotomes (F9/18, Na11/22, Al13/26) are all radioactive and scarce, failing thus to constitute a natural synthesis sequence,

9) During the synthesis process, a deuterotome will only emit beta particles (β^-) if a neutron has been previously captured,

10) The capture of alpha particles is a likely quick way for the synthesis of deuterotomes to occur.

After the atoms of hydrogen (H1/1, D1/2, T1/3) have been formed, the nuclear fusion ensuing between deuterium and tritium and all the other possible ways, such as the proton-proton cycle and the carbon cycle, have become the source of particles (alpha and beta), neutrons, negative electrons, positive electrons (positrons) and neutrinos in a nearly inexhaustible amount, for a very long time. The nuclear fusion takes place when heat is produced, increasing both the temperature and the pressure in the medium. The synthesis process as far as uranium (U92), however, may have happened without the presence of the radioactive property of the elements existing between mercury (Hg80) and uranium (U92). One can reason this way for the synthesis process in that linage is not one of degradation as happens to the four radioactive families (radium, actinium, uranium-235 and thorium).

PICTURE OF THE POSSIBLE WAYS OF DEUTEROTOMES SYNTHESIS

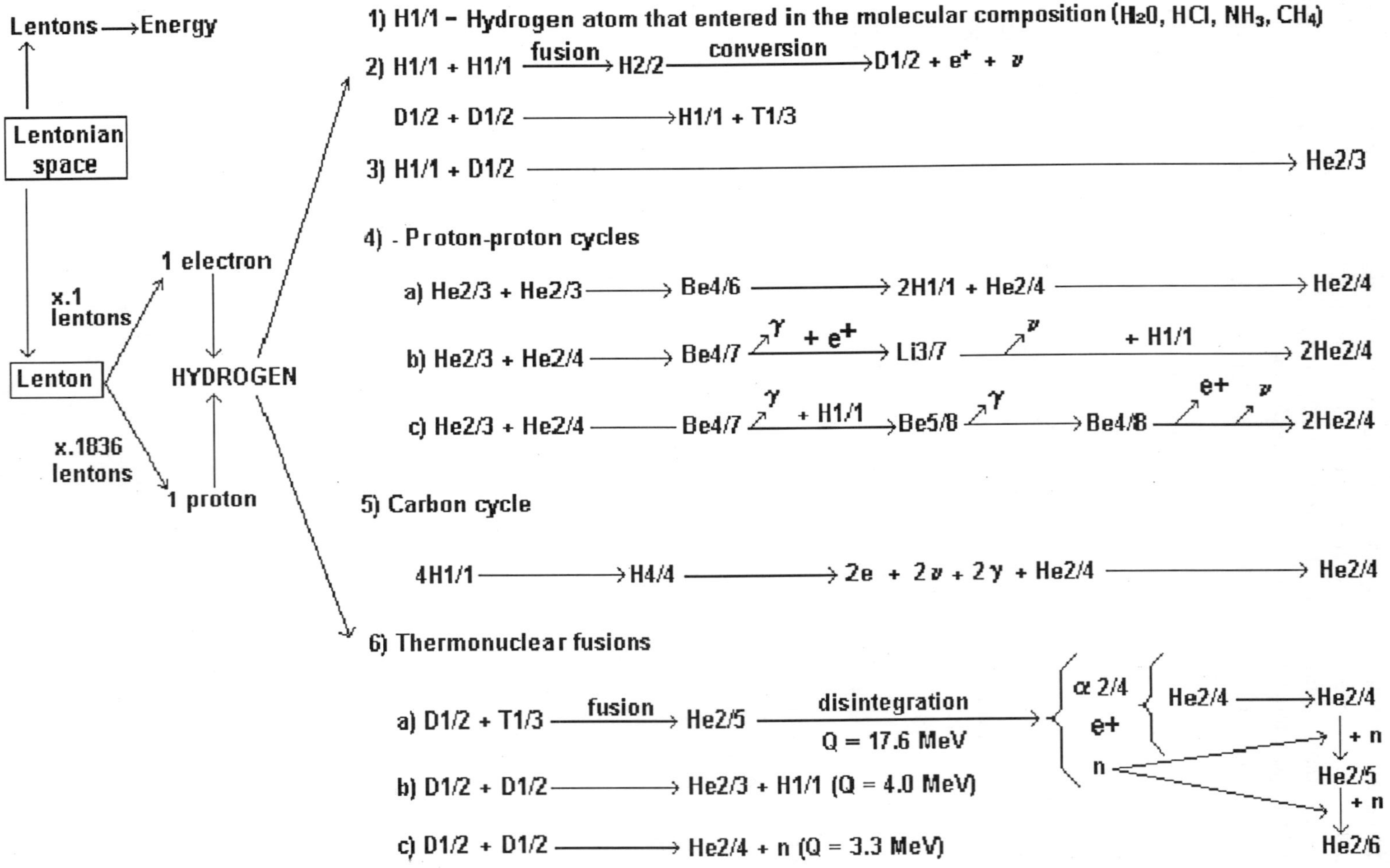

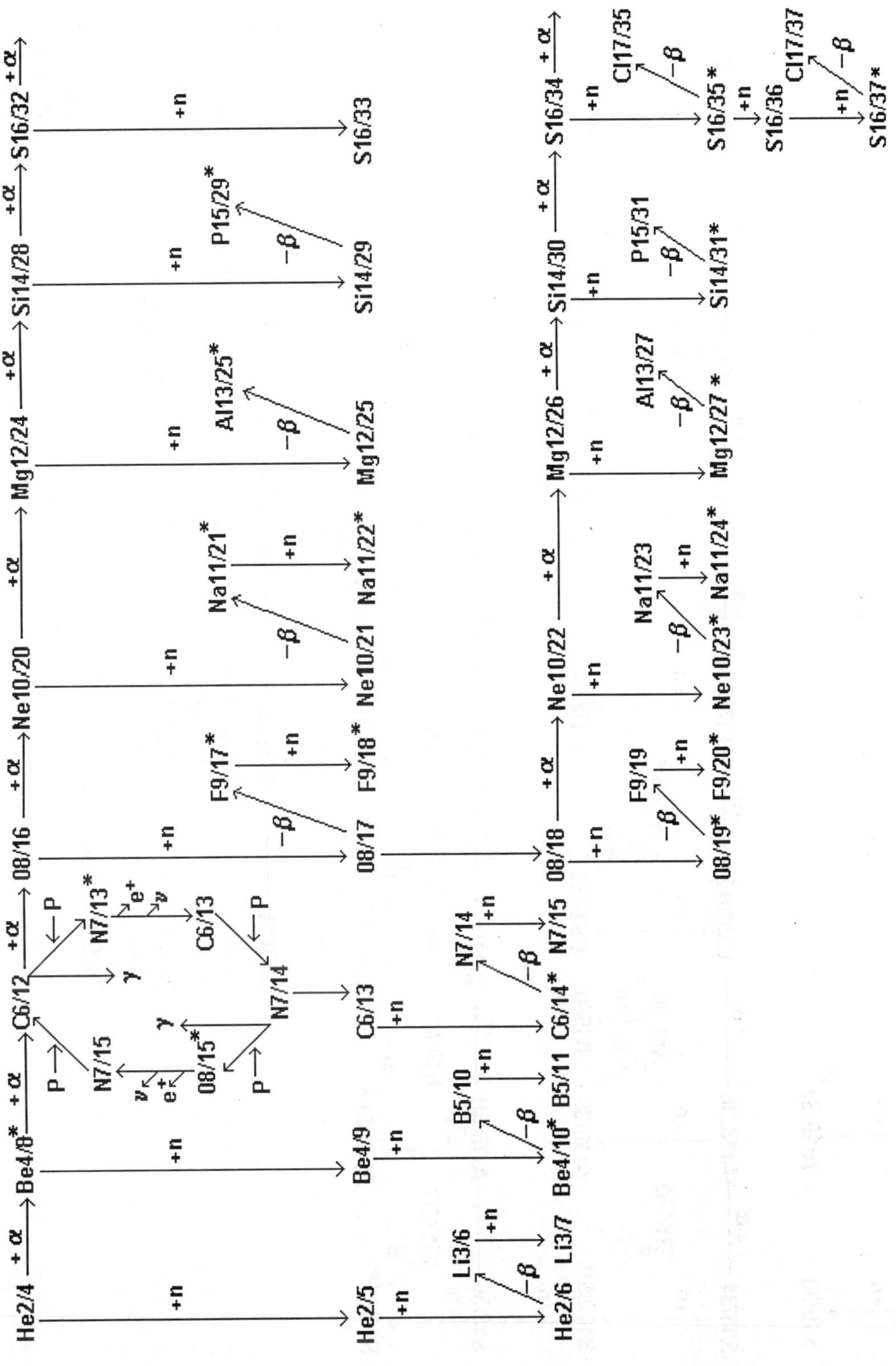
He2/4 + α → Be4/8* + α → C6/12 + α → O8/16 + α → Ne10/20 + α → Mg12/24 + α → Si14/28 + α → S16/32 + α →
+n
He2/5 Be4/9 C6/13 O8/17 Ne10/21 Mg12/25 Si14/29 S16/33
−β
Li3/6 B5/10 N7/14 F9/17* Na11/21* Al13/25* P15/29*
F9/18* Na11/22*
He2/6 Li3/7 Be4/10* B5/11 C6/14* N7/15 O8/18 + α → Ne10/22 + α → Mg12/26 + α → Si14/30 + α → S16/34 + α →
O8/19* F9/19 F9/20* Ne10/23* Na11/23 Na11/24* Mg12/27 * Al13/27 Si14/31* P15/31 S16/35* Cl17/35 S16/36 S16/37* Cl17/37
N7/13* C6/13 N7/14 O8/15* N7/15
P
γ
e+
ν

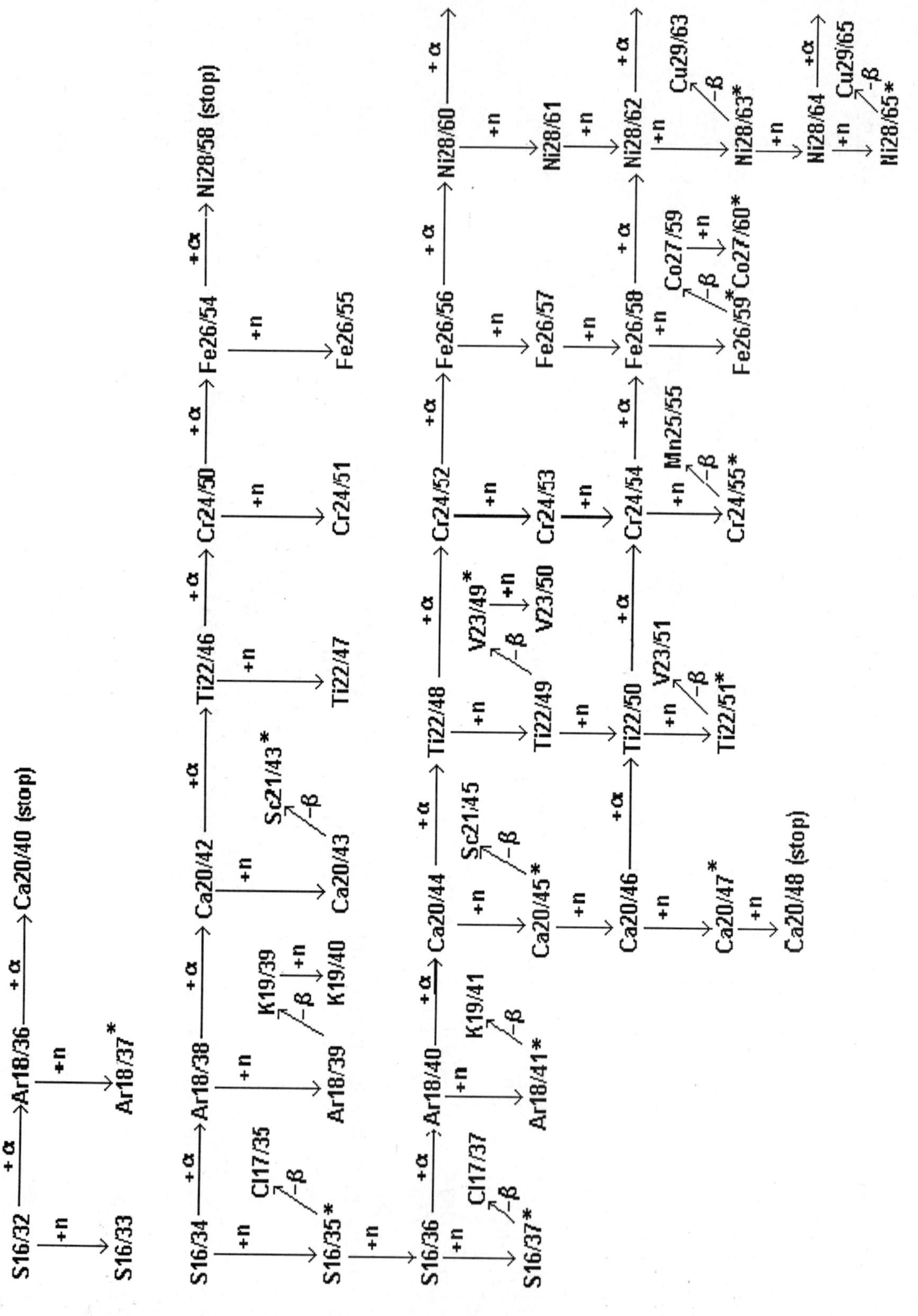
S16/32 +α → Ar18/36 +α → Ca20/40 (stop)
+n S16/33
+n Ar18/37*
S16/34 +α → Ar18/38 +α → Ca20/42 +α → Ti22/46 +α → Cr24/50 +α → Fe26/54 +α → Ni28/58 (stop)
+n S16/35* −β Cl17/35
+n Ar18/39 −β K19/39 +n K19/40
+n Ca20/43 −β Sc21/43*
+n Ti22/47
+n Cr24/51
+n Fe26/55
S16/36 +α → Ar18/40 +α → Ca20/44 +α → Ti22/48 +α → Cr24/52 +α → Fe26/56 +α → Ni28/60 +α
+n S16/37* −β Cl17/37
+n Ar18/41* −β K19/41
+n Ca20/45* −β Sc21/45
+n Ti22/49 −β V23/49* +n V23/50
+n Cr24/53
+n Fe26/57
+n Ni28/61
Ca20/46 +α → Ti22/50 +α → Cr24/54 +α → Fe26/58 +α → Ni28/62 +α
+n Ca20/47* +n Ca20/48 (stop)
+n Ti22/51* −β V23/51
+n Cr24/55* −β Mn25/55
+n Fe26/59* −β Co27/59 +n Co27/60*
+n Ni28/63* −β Cu29/63
+n Ni28/64 +α
+n Ni28/65* -β Cu29/65

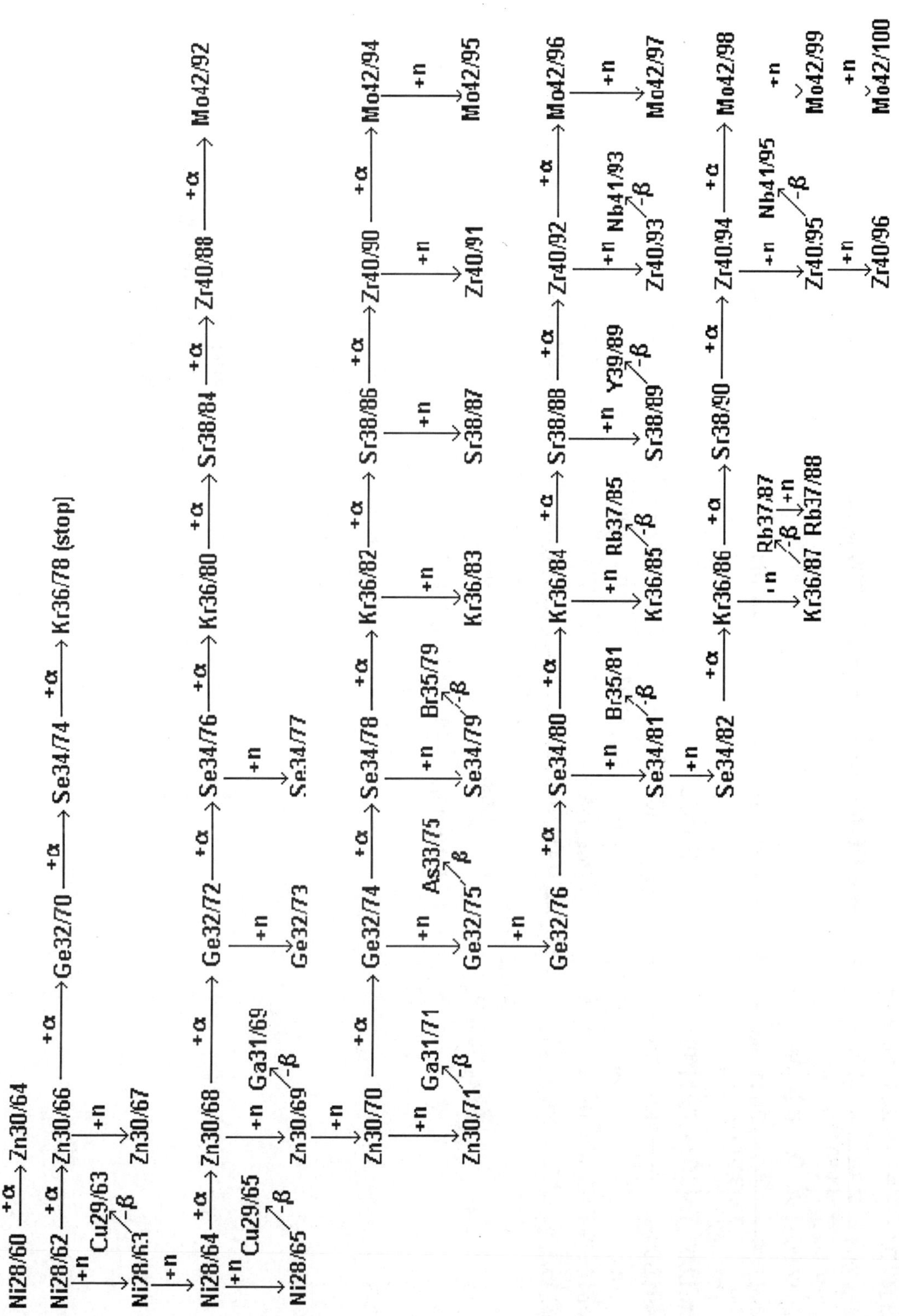
Ni28/60 +α → Zn30/64
Ni28/62 +α → Zn30/66 +α → Ge32/70 +α → Se34/74 +α → Kr36/78 (stop)
+n Cu29/63 -β
Ni28/63 Zn30/67 +n
+n
Ni28/64 +α → Zn30/68 +α → Ge32/72 +α → Se34/76 +α → Kr36/80 +α → Sr38/84 +α → Zr40/88 +α → Mo42/92
+n Cu29/65 -β
Ni28/65 Zn30/69 Ga31/69 -β +n Ge32/73 +n Se34/77 +n
+n
Zn30/70 +α → Ge32/74 +α → Se34/78 +α → Kr36/82 +α → Sr38/86 +α → Zr40/90 +α → Mo42/94
+n Ga31/71 -β Zn30/71
+n As33/75 -β Ge32/75
+n Br35/79 -β Se34/79
+n Kr36/83 +n Sr38/87 +n Zr40/91 +n Mo42/95
+n
Ge32/76 +α → Se34/80 +α → Kr36/84 +α → Sr38/88 +α → Zr40/92 +α → Mo42/96
+n Br35/81 -β Se34/81
+n Rb37/85 -β Kr36/85
+n Y39/89 -β Sr38/89
+n Nb41/93 -β Zr40/93
+n Mo42/97
+n
Se34/82 +α → Kr36/86 +α → Sr38/90 +α → Zr40/94 +α → Mo42/98
+n Rb37/87 -β Kr36/87 +n Rb37/88
+n Nb41/95 -β Zr40/95
+n Zr40/96
+n Mo42/99
+n Mo42/100

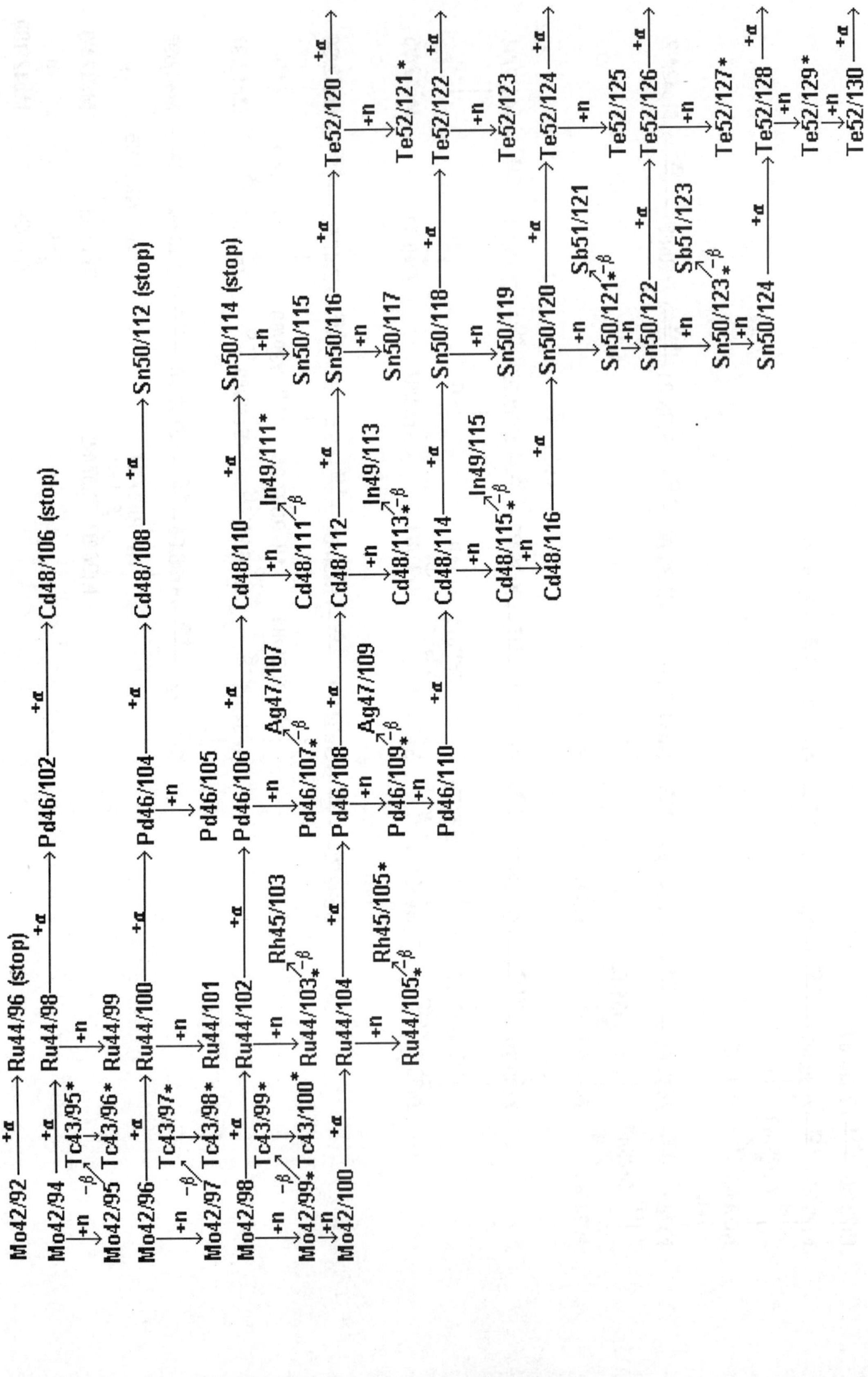
Mo42/92
Ru44/96 (stop)
Mo42/94
Tc43/95*
Mo42/95
Tc43/96*
Ru44/98
Ru44/99
Pd46/102
Cd48/106 (stop)
Mo42/96
Tc43/97*
Mo42/97
Tc43/98*
Ru44/100
Ru44/101
Pd46/104
Pd46/105
Cd48/108
Sn50/112 (stop)
Mo42/98
Tc43/99*
Mo42/99*
Tc43/100*
Ru44/102
Ru44/103*
Rh45/103
Pd46/106
Pd46/107*
Ag47/107
Cd48/110
Cd48/111
In49/111*
Sn50/114 (stop)
Sn50/115
Mo42/100
Ru44/104
Ru44/105*
Rh45/105*
Pd46/108
Pd46/109*
Ag47/109
Pd46/110
Cd48/112
Cd48/113*
In49/113
Cd48/114
Cd48/115*
In49/115
Cd48/116
Sn50/116
Sn50/117
Sn50/118
Sn50/119
Sn50/120
Sn50/121*
Sb51/121
Sn50/122
Sn50/123*
Sb51/123
Sn50/124
Te52/120
Te52/121*
Te52/122
Te52/123
Te52/124
Te52/125
Te52/126
Te52/127*
Te52/128
Te52/129*
Te52/130

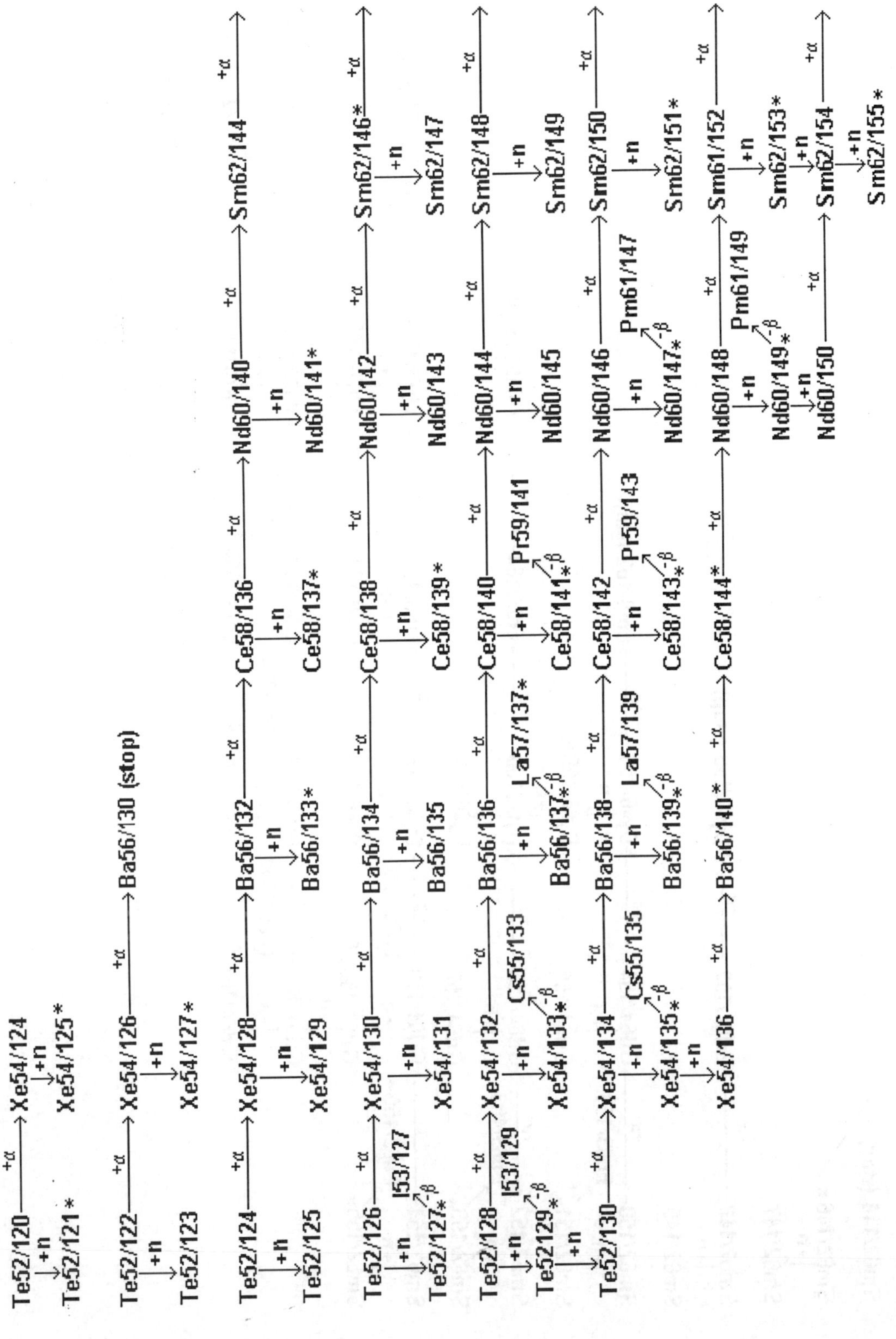
Te52/120 +α Xe54/124
+n Te52/121*
+n Xe54/125*
Te52/122 +α Xe54/126 +α Ba56/130 (stop)
+n Te52/123
+n Xe54/127*
Te52/124 +α Xe54/128 +α Ba56/132 +α Ce58/136 +α Nd60/140 +α Sm62/144 +α
+n Te52/125
+n Xe54/129
+n Ba56/133*
+n Ce58/137*
+n Nd60/141*
Te52/126 +α Xe54/130 +α Ba56/134 +α Ce58/138 +α Nd60/142 +α Sm62/146* +α
+n Te52/127* -β I53/127
+n Xe54/131
+n Ba56/135
+n Ce58/139*
+n Nd60/143
+n Sm62/147
Te52/128 +α Xe54/132 +α Ba56/136 +α Ce58/140 +α Nd60/144 +α Sm62/148 +α
+n Te52129* -β I53/129
+n Xe54/133* -β Cs55/133
+n Ba56/137* -β La57/137*
+n Ce58/141* -β Pr59/141
+n Nd60/145
+n Sm62/149
+n Te52/130 +α Xe54/134 +α Ba56/138 +α Ce58/142 +α Nd60/146 +α Sm62/150 +α
+n Xe54/135* -β Cs55/135
+n Ba56/139* -β La57/139
+n Ce58/143* -β Pr59/143
+n Nd60/147* -β Pm61/147
+n Sm62/151*
+n Xe54/136 +α Ba56/140* +α Ce58/144* +α Nd60/148 +α Sm61/152 +α
+n Nd60/149* -β Pm61/149
+n Sm62/153*
+n Nd60/150 +α Sm62/154 +α
+n Sm62/155*

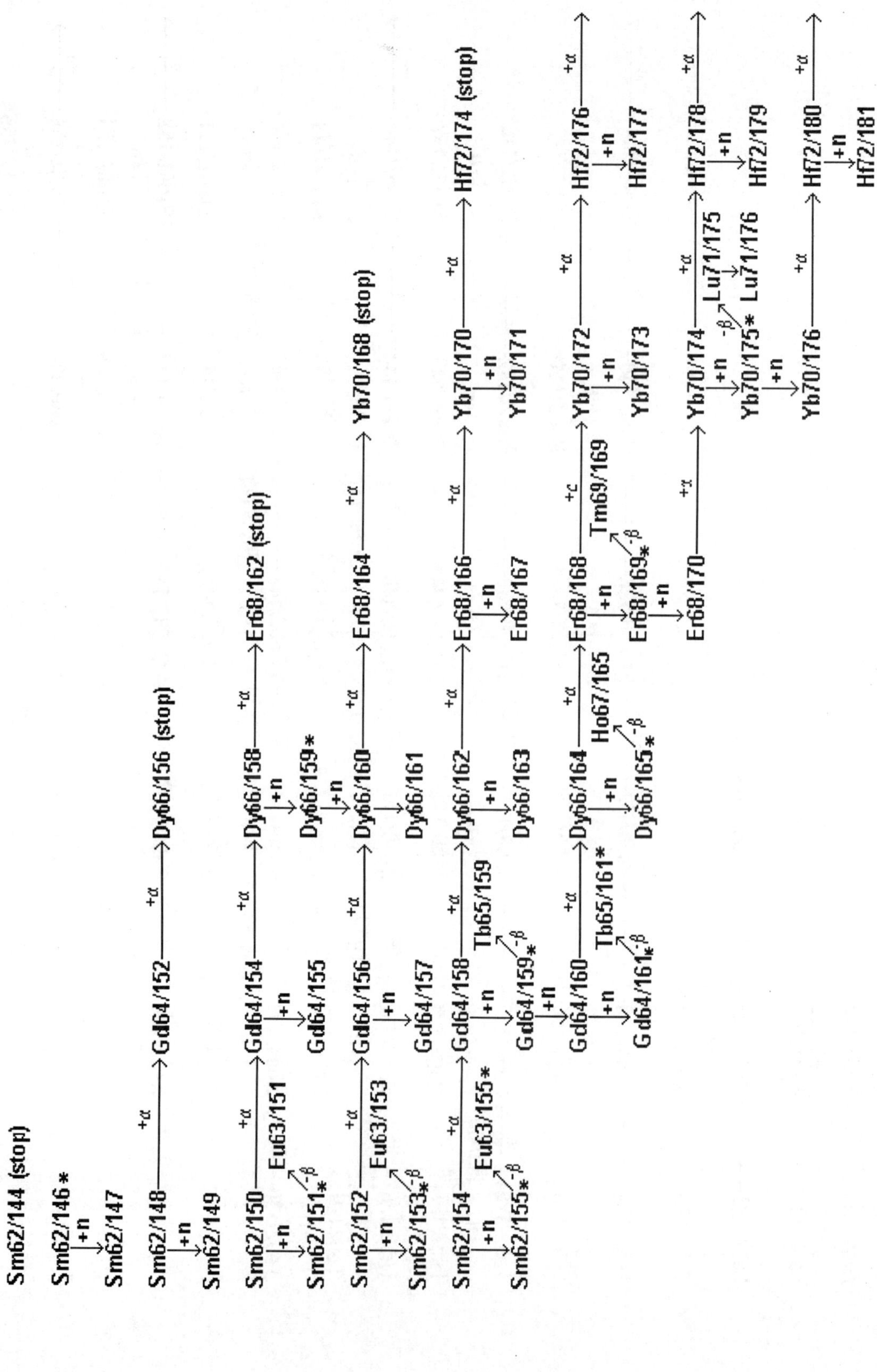
Sm62/144 (stop)
Sm62/146*
+n
Sm62/147
Sm62/148
+α
Gd64/152
+α
Dy66/156 (stop)
Sm62/149
Sm62/150
+α
Gd64/154
+α
Dy66/158
+α
Er68/162 (stop)
Sm62/151*
-β
Eu63/151
Gd64/155
Dy66/159*
Sm62/152
+α
Gd64/156
+α
Dy66/160
+α
Er68/164
+α
Yb70/168 (stop)
Sm62/153*
Eu63/153
Gd64/157
Dy66/161
Sm62/154
+α
Gd64/158
+α
Dy66/162
+α
Er68/166
+α
Yb70/170
+α
Hf72/174 (stop)
Sm62/155*
Eu63/155*
Gd64/159*
Tb65/159
Dy66/163
Er68/167
Yb70/171
Gd64/160
+α
Dy66/164
+α
Er68/168
+α
Yb70/172
+α
Hf72/176
Gd64/161*
Tb65/161*
Dy66/165*
Ho67/165
Er68/169*
Tm69/169
Yb70/173
Hf72/177
Er68/170
+α
Yb70/174
+α
Hf72/178
Yb70/175*
Lu71/175
Lu71/176
Hf72/179
Yb70/176
+α
Hf72/180
Hf72/181

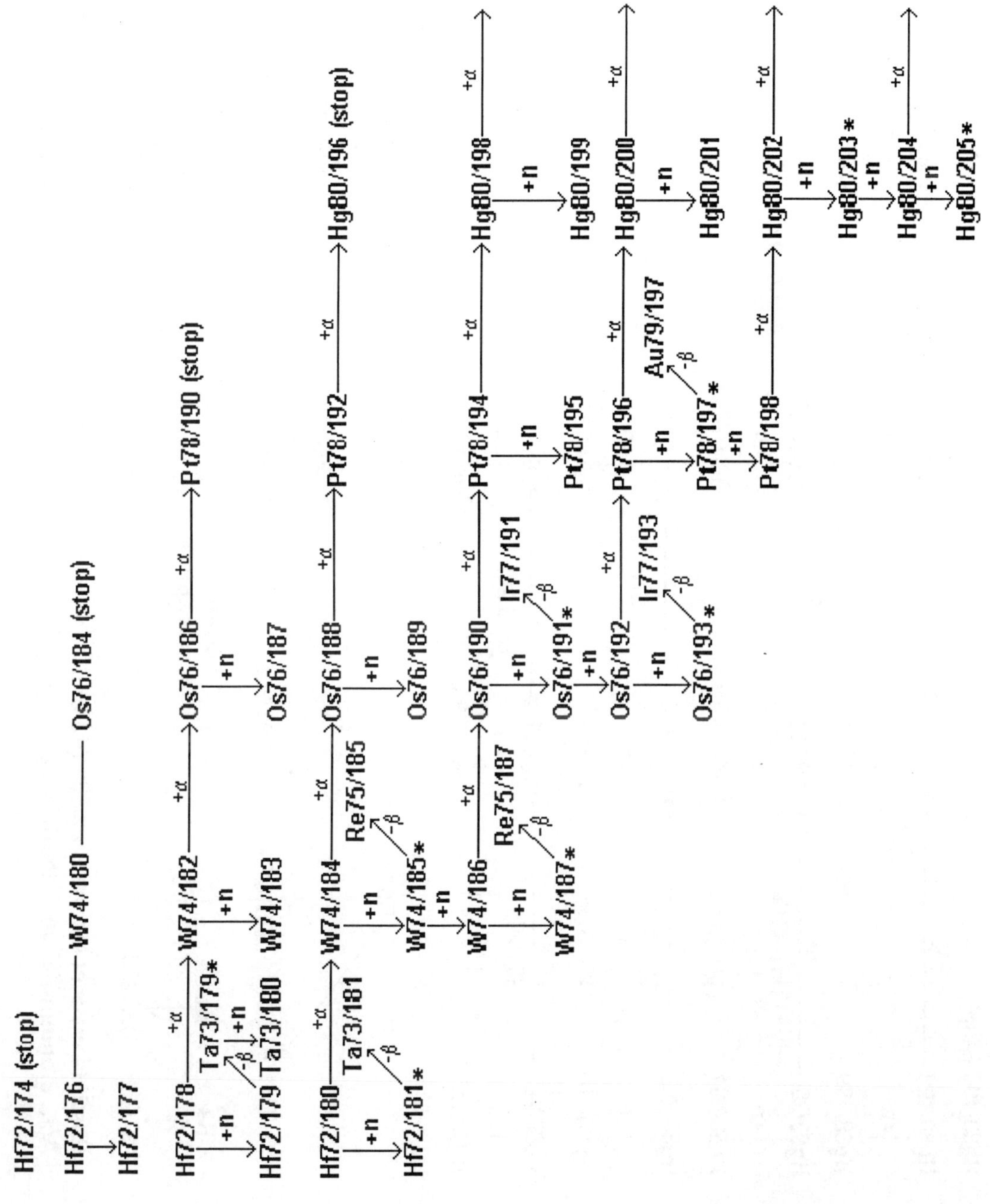
Hf72/174 (stop)
Hf72/176
Hf72/177
W74/180
Os76/184 (stop)
Hf72/178
Hf72/179
Ta73/179*
Ta73/180
W74/182
W74/183
Os76/186
Os76/187
Pt78/190 (stop)
Hf72/180
Hf72/181*
Ta73/181
W74/184
W74/185*
Re75/185
W74/186
W74/187*
Re75/187
Os76/188
Os76/189
Pt78/192
Hg80/196 (stop)
Os76/190
Os76/191*
Ir77/191
Os76/192
Os76/193*
Ir77/193
Pt78/194
Pt78/195
Pt78/196
Pt78/197*
Au79/197
Pt78/198
Hg80/198
Hg80/199
Hg80/200
Hg80/201
Hg80/202
Hg80/203*
Hg80/204
Hg80/205*
+n
+α
-β

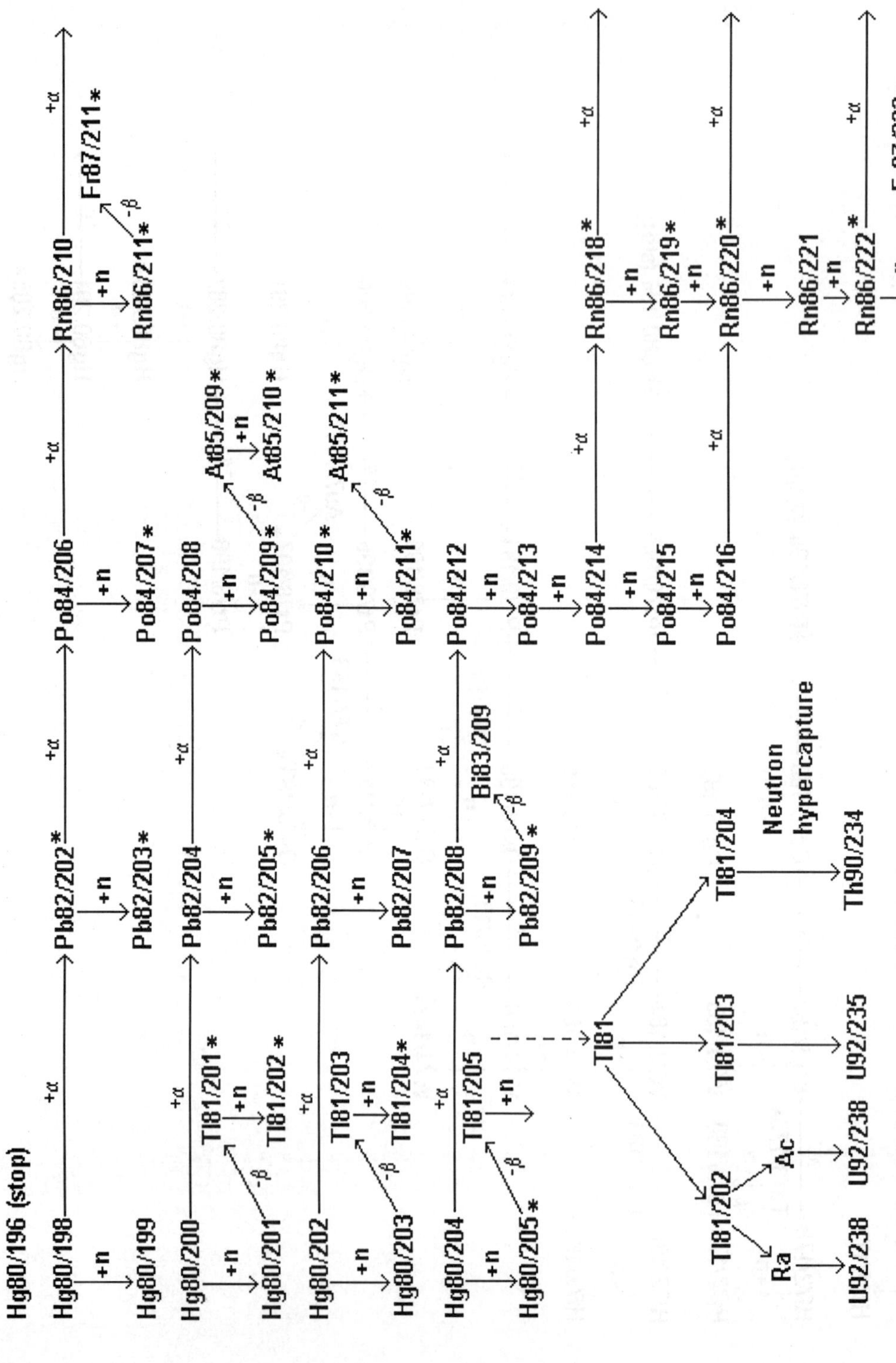

Possible ways of the uranium-238 (Ra and Ac), uranium-235 and thorium synthesis

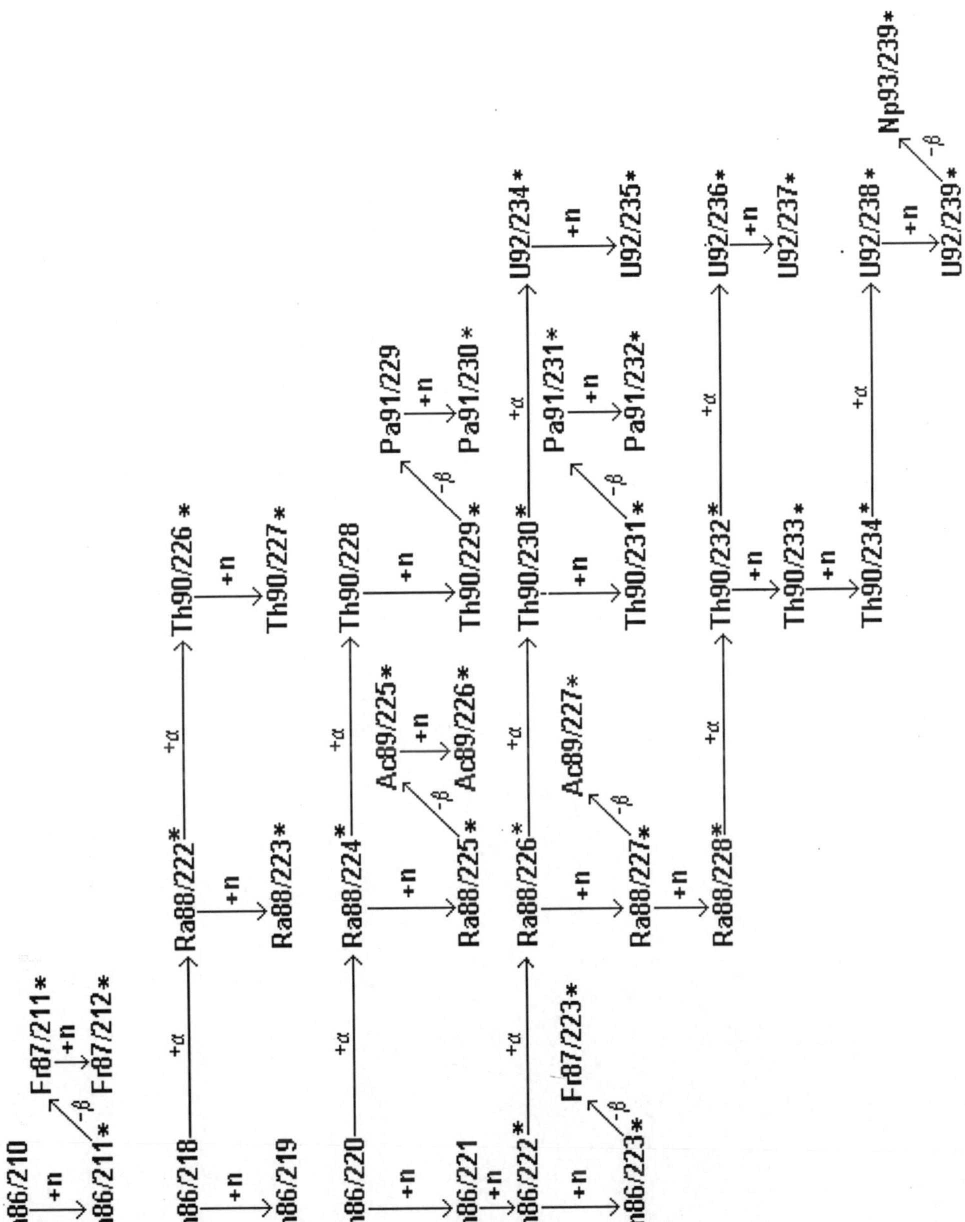
Rn86/210
Rn86/211*
Fr87/211*
Fr87/212*
Rn86/218
Rn86/219
Ra88/222*
Ra88/223*
Th90/226*
Th90/227*
Rn86/220
Rn86/221
Rn86/222*
Rn86/223*
Fr87/223*
Ra88/224*
Ra88/225*
Ac89/225*
Ac89/226*
Th90/228
Th90/229*
Pa91/229
Pa91/230*
Ra88/226*
Ra88/227*
Ac89/227*
Ra88/228*
Th90/230*
Th90/231*
Pa91/231*
Pa91/232*
U92/234*
U92/235*
Th90/232*
Th90/233*
Th90/234*
U92/236*
U92/237*
U92/238*
U92/239*
Np93/239*
+n
+α
-β

CHAPTER 5

Radioactive deuterotomes (light, intermediate and heavy)

5.1 – Isotopes capturing neutrons and emitting beta particles (β): percentage and half-life

Throughout the synthesis process of deuterotomes from helium (He2) to uranium (U92), the mechanism "neutron capture, beta particles emission" goes by a deuterotome that may have two possible characteristics: 1) be radioactive with a short half-life; 2) be stable and found in a small scale among its isotopes. The synthesis sequence of tin (Sn50) and antimony (Sb51) isotopes alike can be cited as an example:

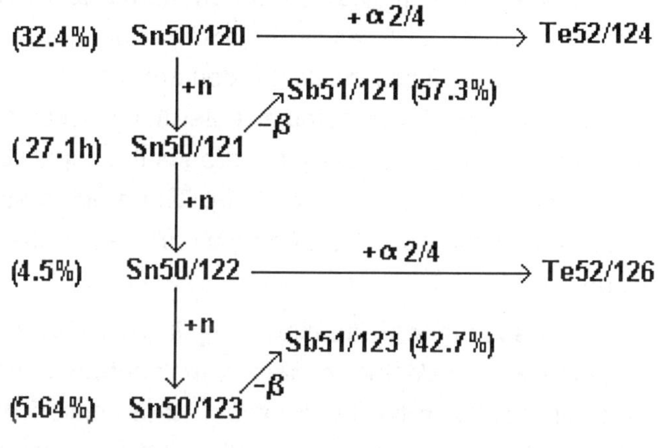

Isotopes stemming from a neutron being captured become radioactive and emit a beta particle (β). Soon after this emission, the resulting deuterotome can be stable. This form of synthesis can be often seen in the "Table of possible pathways for the synthesis of deuterotomes". Below are a few examples.

Neutron captor	+ n	Intermediate isotope	(% and half-life) - β	Final isotope	(% and half-life)
He2/5	----------------- >	He2/6	(extinct)------------- >	Li3/6	(7.5%)
Be4/9	----------------- >	Be4/10	1,6 My -------------- >	B5/10	(19.8%)
C6/13	----------------- >	C6/14	5,730 y ------------ >	N7/14	(99.63%)
O8/18	----------------- >	O8/19	26.9 s --------------- >	F9/19	(100%)
Ne10/22	--------------- >	Ne10/23	37.6 s --------------- >	Na11/23	(100%)

Ca20/44 --------------	> Ca20/45	165 d ---------------	> Se21/45	(100%)
Fe26/58 --------------	> Fe26/59	446 d ---------------	> Co27/59	(100%)
Ge32/72 --------------	> Ge32/73	7.8% ---------------	> As33/73	(80.3 d)
Kr36/86 ---------------	> Kr36/87	76 m ----------------	> Rb37/87	(27.83%)
Ru44/102 ------------	> Ru44/103	39.4 d ---------------	> Ro45/103	(100%)
Hg80/202 ------------	> Hg80/203	46.8 d ---------------	> Tl81/203	(29.5%)
Pb82/208 ------------	> Pb82/209	3.25 h ---------------	> Bi83/209	(100%)

Isotopes originating from this mechanism are mostly radioactive with a variable half-life of a few seconds to some thousand years. Isotopes resulting from beta particles being emitted are mostly stable and only a few are radioactive.

In order for a deuterotome to be able to create an isotope with an immediately higher mass number, it must necessarily be located in an atmosphere where neutrons are actually emitted; in other words, it depends on a source of neutrons. Therefore, the number of isotopes formed via that mechanism is dependent on the number of neutrons supplied. Conversely, the emission of beta particles (β) it not dependent on any external means, for the beta decay process takes place inside the deuterotome.

Because this mechanism produces a stable deuterotome, it can be taken for granted that the capture of neutrons, with the consequent drop in the capacitance value, causes the deuterotome then formed to become unstable. The electric stability, in terms of capacitance, is restored upon the emission of a beta particle and the consequent rise in the capacitance value.

Let's consider the capacitance equation: **C = k. A. np/d. nn,** where **C** is the deuterotome capacitance; **k** the dielectric constant of the neutron; **A** the area of the proton plate; **np** the number of protons in the deuterotome; **d** the distance among the plates (protons); and **nn** the number of neutrons participating in the deuterotome's composition. 1eF.p-1 expresses the k value and **A** has a constant value 1.5708 p^2; **d** is equal to 1p (diameter of a neutron). The capacitance values of elements formed during the synthesis process can be then calculated as per the following equation:

Mg12/26 + n ————> Mg12/27 – β ————> Al13/27

By utilizing the first equation mentioned above, the capacitance values will be as follows:

Mg12/26 C = 1.3464 eF (12 protons + 14 neutrons)

Mg12/27 C = 1.2566 eF (12 protons + 15 neutrons)

Al13/27 C = 1.4586 eF (13 protons + 14 neutrons)

Below is a chart showing the capacitance value in each deuterotome associated with the respective number of neutrons.

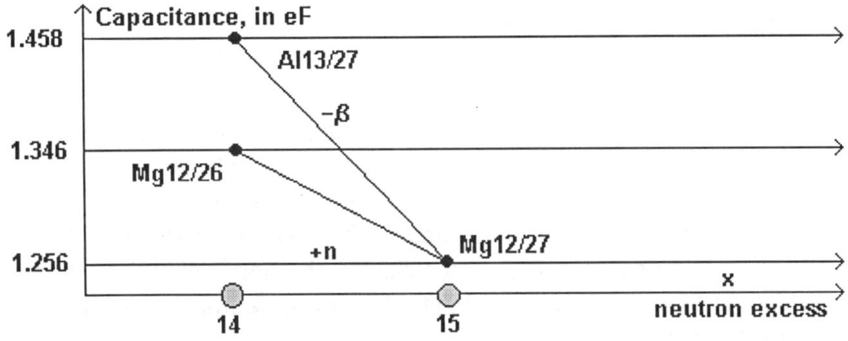

Fig. 31 – Variations in the capacitance values of deuterotomes in the synthesis mechanism "neutron capture, emission of beta particles" is graphically represented. The numbers on axis x represent the excess of neutrons beyond the number of protons. The capture of neutrons reduces the capacitance value while the emission of beta particles increases its value.

The capture of neutrons is the only synthesis process capable of reducing the capacitance value. The emission of beta particles and the capture of alpha particles alike increase its value. Light deuterotomes emitting beta particles like magnesium-27 (Mg12/27) are radioactive as are intermediate deuterotomes like zirconium-95 (Zr40/95) and heavy ones like polonium-211 (Po84/211).

Consequently, radioactivity by beta emission is a property that should not be based only on the so-called "excessive number of neutrons" in relation to the number of protons in heavy elements, where the values in the neutron/proton ratio are above 1.5 such as uranium-238 (U92/238) where the excess of neutrons is 146 and the number of protons is 92 with a 1.58 ratio.

5.2 – Conditions determining radioactivity

The value variation in the total capacitance of deuterotomes in keeping with Fig. 31, in elements capturing neutrons, can be an important factor leading to the radioactivity in light elements like magnesium-27 (Mg12/27), for example. In heavy deuterotomes, though, the radioactive phenomenon can be dependent on several factors, such as:

1) - Instability in the capacitance values of deuterotomes;

2) - Capacitance of the alpha-unit that makes up the "alpha-appendix";

3) - The electric load in the "electronic furnace";

4) - Lower binding energy per nucleon in heavy deuterotomes.

These factors can play an important role, either individually or by combining their forces. In addition to the factors referred to above, other conditions participating in the process of radioactive degradation can be cited:

1) - The radioactive degradation from uranium (U92) to thallium (Tl81), passing by radon (Rn86), is an evidence that this noble gas did not go through the physical volume-reduction phenomenon; therefore, its structure is not armored, as can be otherwise seen in other noble gases,

2) - The value in the mass number of each alpha-unit is a multiple of four in the natural series consisting of whole numbers as follows: α2/4, α2/8, α2/12; in the radioactive families of both radium and actinium, these three multiples are present; in the family of thorium only the alpha-units α2/4 and α2/8 are encountered while in the radioactive family of U92/235 there are two alpha-units α2/6, rather than the alpha-unit α2/12,

3) - The process of radioactive degradation of heavy transuranium deuterotomes occurs in the fraction of these deuterotomes called "alpha appendix"; the alpha appendix of thorium (Th90) has five alpha-units and the other three deuterotomes (U92/238, U92/238 and U92/235) have six,

4) - The capacitive bindings in the alpha-units of the last six protons of uranium (U92), when present, are parallel-equivalent and their electrons do not take part in the bundle of 32 electrons making up the "furnace",

5) - In thallium deuterotome (Tl81), the number of electrons in the "electronic furnace" is reduced from 32 to 27, consequently reducing its electric load and power over the bindings or proton-neutron junctions, and becomes insufficient to break those junctions; as a result of the reduction in the "furnace" electric load, the deuterotomes then stop emitting alpha particles (α2/4) and the mechanism of radioactive degradation type "emission of alpha particles (α2/4) and emission of beta particles (β)" is consequently ceased.

5.3 – Possible ways for the nucleus-synthesis of uranium (U92/235, U92/238) and thorium (Th90/232) deuterotomes from thallium (Tl81/203, 204 and 205)

The nucleus-synthesis in both uranium and thorium can be analyzed from thallium (Tl81). The radioactive families of radium and actinium originate in thallium-203; the radioactive family of uranium-235 originates in thallium-204 and the one of thorium stems from thallium-205. The schemes below show the possible synthesis ways by the process of neutrons hypercapture and emission of beta particles (β), passing by well-known isotopes previously identified and by others not specifically referred to, which may have existed. Several intermediate isotopes have a very short half-life; for this reason they are no longer encountered in nature or are scarcely available.

The reactions illustrated below show a possible synthesis way in the "alpha appendix" from mercury (Hg80) isotope with a phylogenetic code in its structures. The sequence of neutrons following each proton contains a phylogenetic code that will possibly determine the sequence in the emissions of alpha (α) and beta particles (β) in each radioactive family. The sequences referred to in the nucleus-synthesis process apparently seems to be the only ones capable of recording the code of radioactive degradation in the "alpha appendix" of each radioactive family.

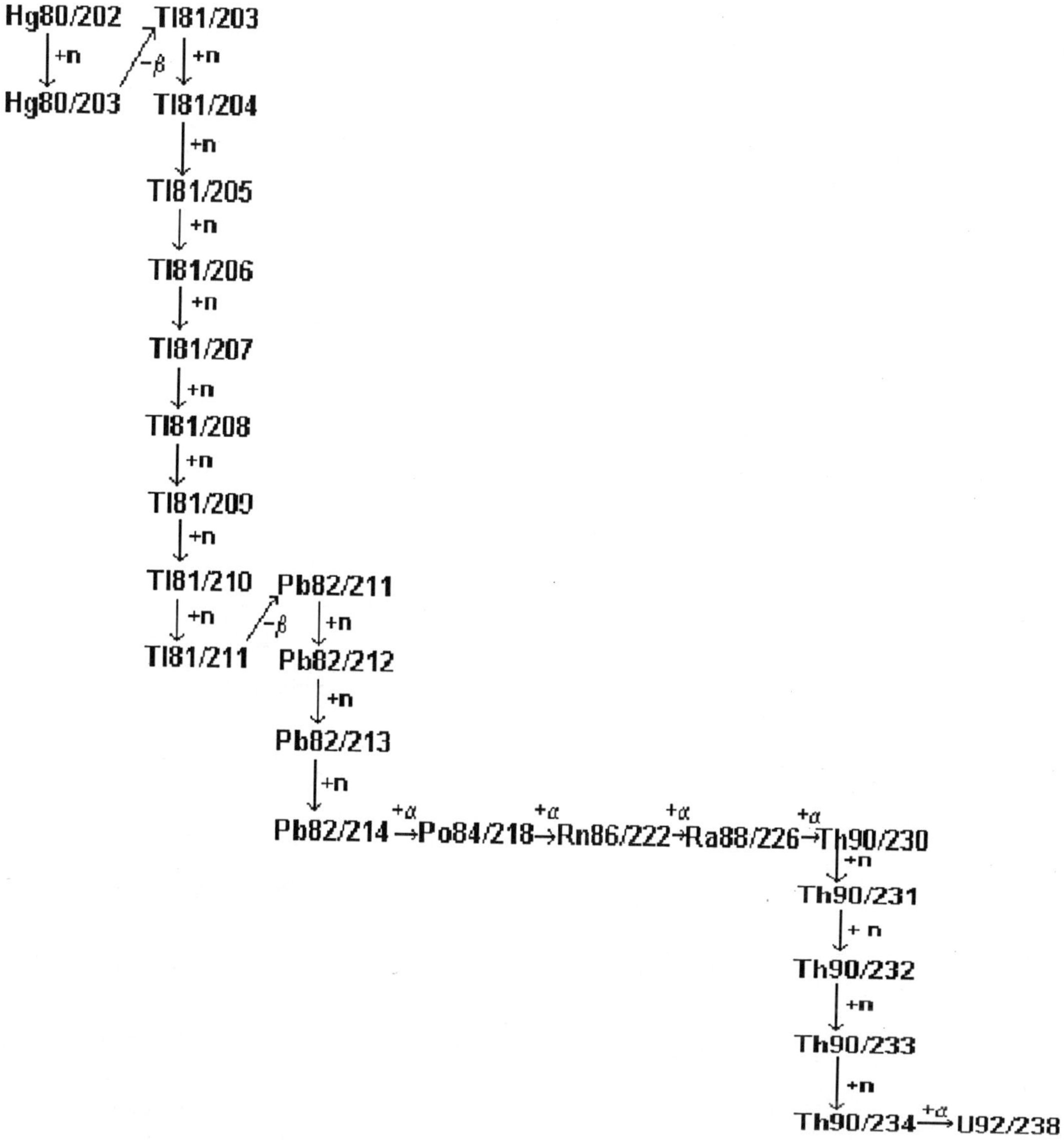

Fig. 32. Possible ways of nucleus-synthesis in uranium-238 (U92/238) in the radioactive family of radium from thallium-203 (Tl81/203) by way of neutrons hypercapture, beta particles emission (β) and alpha particles capture (α2/4).

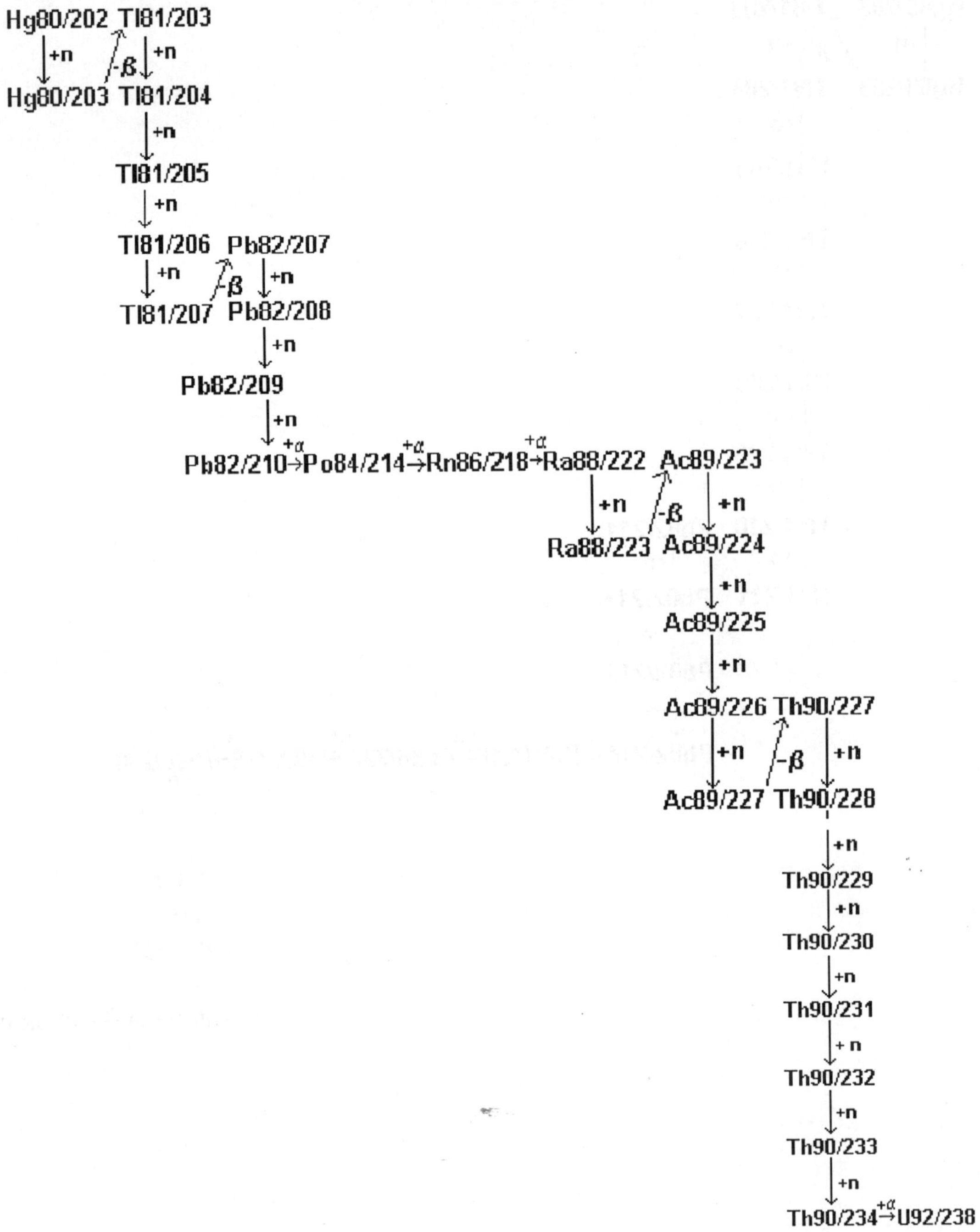

Fig. 33. Possible ways of nucleus-synthesis in uranium-238 (U92/238) in the radioactive family of actinium from thallium-203 (Tl81/203) by way of the following mechanisms: neutrons hypercapture, beta particles emission (α) and alpha particles capture (α2/4).

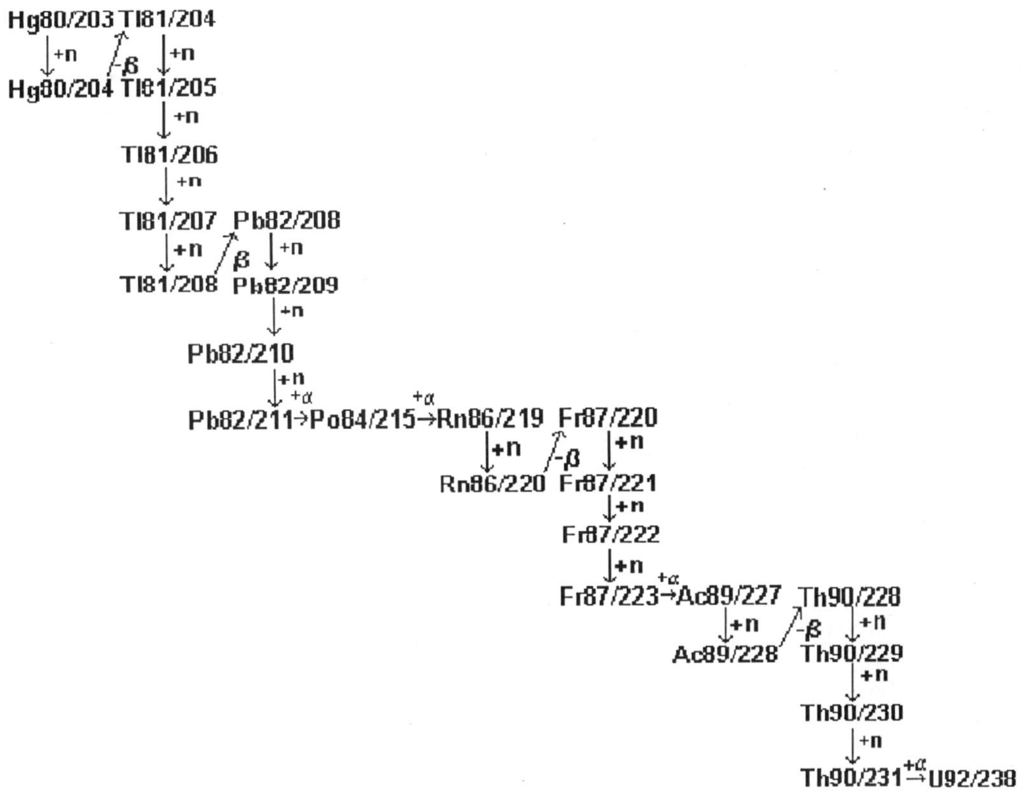

Fig 34. Possible ways of nucleus-synthesis in uranium-235 from thallium-204 (Tl81/204) by way of neutrons hypercapture, beta particles emission (β) and alpha particles capture.

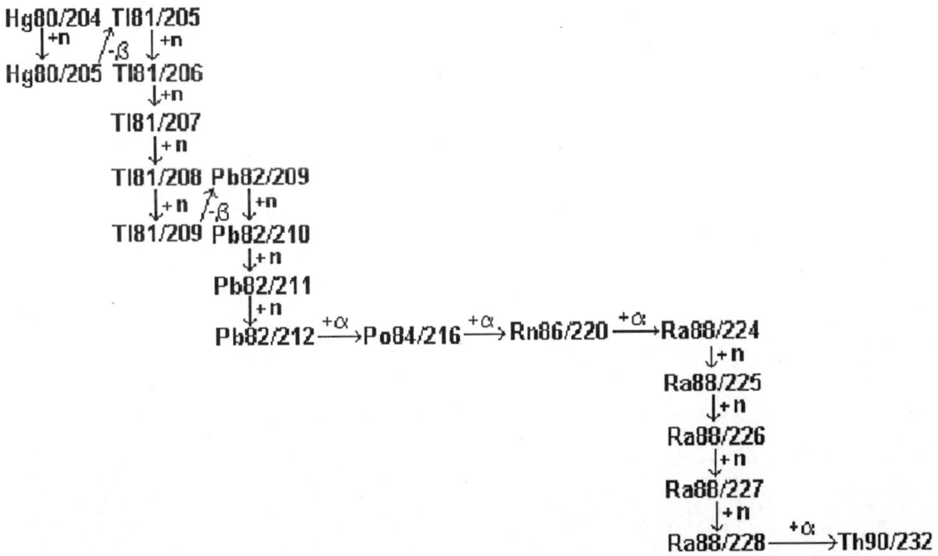

Fig 35. Possible way of nucleus-synthesis in thorium-232 from thallium-205 (Tl81/205) by way of neutrons hypercapture, beta particles emission (β) and alpha particles capture.

CHAPTER 6

Noble gases: physical phenomenon of volume-reduction

6.1 – Avogadro's number

According to Avogadro and Ampère (mathematician and French physicist, 1775-1836), "similar volumes of any gas at identical temperature and pressure (0ºC and 760 mmHg) contain the same number of molecules". Thus, in compliance with the experiments previously carried out, it was determined that a molecule-gram of any gas occupies the volume of 22.4 liters.

Avogadro's number (Italian physicist, 1776-1856) is a constant value, equal to 6.06 x 10^{23} molecules of a gas occupying a volume of 22.4 liters. This number is represented by **N** for mono-deuterotomic gases (He, Ne, Ar, Kr, Xe), by **2N** for bi-deuterotomic (Cl_2, Br_2, H_2) and by **3N** for tri-deuterotomic (0_3) molecules.

Loschmidt's number (Josef Loschmidt, Austrian physicist, 1821-1895) is the number of molecules of a gas in a cubic centimeter. Hence, this value corresponds to Avogadro's number divided by a volume of 22.400 cm³ (or 22.4 L), whose value is 26 x 10^{18} molecules/cm³.

A number of experiments carried out by various researchers pointed out that noble gases have high density. This property must stem from the physical phenomenon of volume-reduction that noble gases undergo when the electric loads of their pairs of deuterons are reciprocally neutralized (series-equivalent connected) and with the electromotive force of deuterons reduced to 1 eV. Thus, the volume of a noble gas deuterotome (He2, Kr36 and Xe54) is identical to that of light hydrogen (H1/1), that is, 1. The deuterotomes of neon (Ne10) and argon (Ar18) noble gases have higher volumes, twice as much the volume of light hydrogen, which is equal to 2. Their individual density, expressed in g/cm³, is as follows: He = 4, Ne = 10.1, Ar = 19.97, Kr = 83.7 and Xe = 131.3. Fig 36 shows a deuterotome of helium-4 with series-equivalent bindings before going through the physical phenomenon of volume-reduction (on the left), with the volume of the atom of light hydrogen previously reduced.

6.2 – Noble gases: factors leading to the volumetric reduction

According to this new deuterotomic theory now proposed, Avogadro's number is so valuable and so true that it can be considered the cornerstone that leads to the knowledge that accounts for the volumetric reduction of noble gases, because they acquire an extremely high density, far beyond the values encountered in high-density metals such as gold (19.3 g/cm³), mercury (13.59 g/cm³), uranium (18.7 g/cm³), platinum (21.45 g/cm³) etc. Xenon (Xe54) density, for example, is 131.3 g/cm³. This high density does not result from the compression of molecules. The physical phenomenon of volume-reduction that ensues upon the neutralization of noble gas molecules is related to the characteristics of capacitors of fixed capacitance on a given condition and variable on another. Thus, the volumetric reduction of a noble gas causes its density and capacity to increase and confers upon it the features of the genetic pole of a deuterotome.

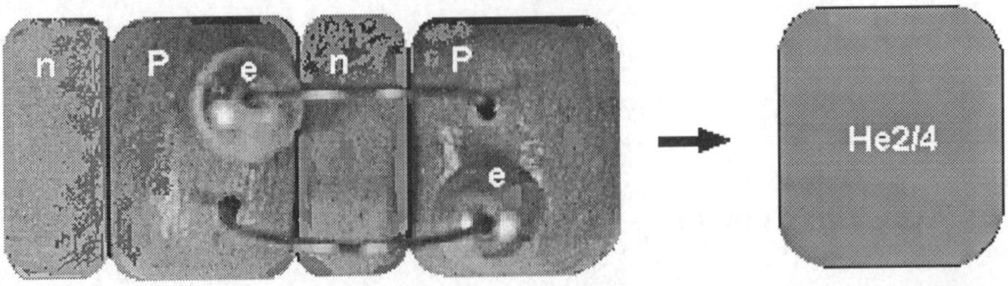

Fig. 36 - **Left**: illustration of a deuterotome of helium-4 (He2/4) noble gas going through reciprocal neutralization of the pairs of deuterons, via series-equivalent bindings which turns it into an inert gas, that is, without any free valence for chemical reactions. **P** is the proton, **n** the neutron and **e,** the electron. In the aftermath of neutralization–the electromotive force of each deuterotome becomes equal to 1eV–the physical phenomenon of volume-reduction occurs and the deuterotome now has a volume similar to that of light hydrogen atom (**right**).

According to this new theoretical proposal, the identification of likely mechanisms for the synthesis of deuterotomes by "alpha particle capture", "neutron capture" and "neutron capture and beta particle emission" types, in addition to the conception of deuterium and tritium as single atomic capacitors gives rise to the notion that a noble gas deuterotome works as the genetic pole to another deuterotome and further, that an increment of deuterons to a noble gas–through synthesis mechanisms–constitute the structure of that deuterotome. Thus, the valences available in a certain element are not in excess of seven, pointing out that all the other deuterons–those which fail to be free valences–must necessarily be electrically neutralized so that chemical reactions may take place in keeping with a pattern or in compliance with a law for chemical combinations. The capacitive property of any noble gas has led to the notion as to the existence of high electric load on the first post-noble gas deuterium, its value being expressed by electron-volt (eV) and the progressive reduction of its electric load per each deuterium that forms the series of the electronic shell of noble gases up to xenon (Xe54). Because the pairs of deuterons that make up the series are neutralized through series-equivalent bindings, this neutralization produces the physical phenomenon of volume-reduction. It is due to the reciprocal neutralization of the pairs of deuterons plus the volume-reduction that noble gases acquire a volume identical to that of light hydrogen atom or twice as much in the case of noble gases neon and argon–starting to have high density. The illustration below shows a neon deuterotome (Ne10) in its linear shape before it undergoes a volume-reduction.

Fig. 37 – Illustration of the deuterotome structure of noble gas neon (Ne10) undergoing total neutralization before the physical phenomenon of volume-reduction has happened. The genetic pole of neon is formed by noble gas helium-4 (He2/4). Deuterons from D3 to D10 complement the deuterotome structure. The electric loads of deuterons are neutralized in relation two their respective pairs by way of series-equivalent bindings. Generally, the deuterotome is electrically neutral, that is, its deuterons have an electromotive force of about 1 eV each. In the next phase, the entire structure goes through a volumetric reduction and its volume becomes twice the volume of light hydrogen. This phenomenon happens to all the noble gases (He2, Ne10, Ar10, Kr36 and Xe54).

The volumetric reduction is the last phenomenon that sets up a noble gas deuterotome within Avogadro's number; otherwise, the number of noble gas deuterotomes per cubic centimeters would be smaller (Loschmidt's number) if heavier noble gases–e.g. xenon–were taken as the basis for the calculation.

6.3 – Physical phenomenon of volume-reduction: high density of noble gases and increase of capacitances

When two industrial capacitors are series-connected, the total capacitance of the system will be lower than the lowest individual capacitance. Two capacitors C1 and C2, whose capacitances rates are 0.15 eF and 0.25 eF, respectively, their total capacitance can be calculated by the equation: Ct = C1 x C2 / C1 + C2. By applying the rules above to the equation, one obtains 0.0625eF, therefore, lower than C1 value. For the calculation of three or more capacitors, this equation cannot be used.

Thus, in order for the capacitance rate to be increased, the conditions cited below are necessary:

1) – Increase area (**A**) of the plates and increase the number of plates,
2) – Reduce distance **d** among the plates, that is, reduce the thickness of the dielectric,
3) – Use a dielectric with a dielectric constant **k** with a higher value.

In the single atomic capacitor (deuterium), the three variables above do not go through any sort of modification as to their values and is therefore, a **fixed capacitor**. However, the volumetric reduction of noble gases causes the thickness of dielectrics (neutrons) to be reduced, increasing the capacitance rate and turning them into **variable capacitors**. The thickness reduction of the protons (plates) does not interfere with the capacitance, for area A is not altered.

If one takes into account the various noble gases with growing deuterotomic numbers–He2, Ne10, Ar18, Kr36, Xe54–each one of them will also have a growing number of plates, which increases their individual capacitance rate. Thus, helium-4 for example consists of a capacitor with two receptor plates whose capacitance value (**C**) equals 3.1416eF; xenon has 54 receptor plates (**multiplate capacitor**) and has a far greater capacitance of 84.8232eF. The photos below show two industrial capacitors series-connected (top of photo) and the figure of helium-4 (He2/4) deuterotome with its electric loads reciprocally neutralized, connected through series-equivalent bindings.

Fig. 38. **Below**: illustration of helium-4 (He2/4) deuterotome going through reciprocal neutralization of the deuterons, prior to the physical phenomenon of volume-reduction. The bindings are series-equivalent. **Above**: two industrial capacitors with the same individual capacitance ± 10 eF, equal to 250 VAC. If the capacitors are series-connected and submitted to spark formation, the one with greater electric load will charge the other. The total capacitance will be lower than the lowest individual capacitance.

6.4 – Noble gases as deuterotomes and genetic poles

Both the physical and chemical properties of deuterotomes are conditioned by the characteristics of their genetic and chemical poles, as well as by the number of deuterons that form a series of electronic layer, in addition to the electromotive force (emf) of each deuterium atom. Consequently, deuterons D9 and D17, respectively of fluorine (F9) and chlorine (Cl17) deuterotomes have identical emf value (1.14 eV) plus the same number of deuterium atoms that make up each individual series. However, the noble gas responsible for the formation of the genetic pole in each series–L and M–is different both in the capacitance and volume. Helium (He2/4) has volume 1, similar to that of light hydrogen atom, but neon (Ne10) has volume 2, twice the volume of helium. Helium deuterotome, as the genetic pole, confers upon the first deuterotome in the series of L-shell, that is lithium (Li3), with an electromotive force of 8 eV.

Neon (Ne10) also confers the same value of electromotive force upon the first deuterotome in its series (M) deuterotomes–sodium (Na11/23). Accordingly, two deuterotomes can possess an identical emf such as oxygen (O8) and sulfur (S16) whose electromotive force equals 1.33 eV. Nitrogen (N7) and phosphorus (P15) have an emf of 1.66 eV. However, their individual genetic pole is different (a) in volume; (b) in density. Helium has a density of 4 g/cm³, while neon density is 10.1 g/cm³; and (c) in capacitance. The facts mentioned above are important for organic chemistry because in a chemical-biological reaction a certain molecule will not capture a sulfur (S16) deuterotome instead of oxygen (O8) simply because they are both bivalent with similar emf in the deuterons of valences. The capacitance of the genetic poles can contribute towards the selection of the deuterotome, given that helium-4 capacitance is 3.1416 eF and that of neon 7.854 eF.

The same thing happens to the deuterotomes of bromine (Br35) and iodine (I53), which have an identical electromotive force of 1.05 eV. However, the genetic pole of each has a different capacitance: the genetic pole of the former is argon (Ar18) with 14.1372eF whereas the genetic pole of the latter is krypton (Kr36) with 56.548eF. The capacitance of these two halogens is also different: bromine rates 1.2495eF, iodine 1.1250eF. Argon has volume 2 and density of 19.97 g/cm^3, while krypton has volume 1 and density of 83.7 g/cm^3. The two halogens also have the same number of post-noble gas deuterons, that is, 17 deuterons.

If a new series of deuterotomes sets out with 18 eV (electric load) for those connected to the N-shell and O-shell, each new deuterotome then formed this load will be successively divided by 2, 3 etc up to 17; when divided by 18, the electric load will equal 1 eV. Therefore, the entire structure becomes electrically neutral, undergoes a volumetric reduction and reaches the volume of light hydrogen atom. With the electric load in a neutral condition, the deuterotome is devoid of free valences for chemical reactions to occur.

The proton is the capacitor plate where electrons are stored. This energy is a reactive electric charge, in other words, the one that carries out the chemical reaction among the atoms, among the deuterons and among atoms and deuterons. The rate of reactive electric charge is given by the distance separating the proton from the electron (proton-electron distance). This distance is progressively equal to 1 for the K-shell, 4 for L and M-shells, 9 for N and O-shells and 16 for P and Q-shells. Each of these distances corresponds or is equal to the resistance that the electric current should overcome between the proton and the electron expressed in electron-ohm (eΩ). The potential difference (ΔV) of the initial load in each deuterium that combines with the noble gas of the genetic pole is the resistance value (R) multiplied by intensity (i), in eA (electron-Ampere) of the electric current. So the potential difference (ΔV) for the series of shells is as follows: 2 eV for K; 8 eV for L and M; 18 eV for N and O, and 32 eV for P and Q. For the protons associated with the K-shell series, the basic electric load is 2 eV. For the remaining series of shells, from L to Q, the load value is equal to the product of the basic load (2 eV) multiplied by the distance between the proton and the electron, that is, 4 (L and M), 9 (N and O) and 16 (P and Q).

Every deuterium belonging to a given series that is added to the previous one has its electric load reduced. This reduction rate will be by half, one third, one fourth, one fifth, one sixth or one seventh should it be an element related to the L or M series. In the case of elements related to the N or O-shell series, the load rate will be the result of the division of 18eV by the number of post-noble gas deuterons that make up the series, for instance, 18eV/1 for potassium (K19), 18eV/2 for calcium (Ca20), 18eV/3 for aluminum (Al13) and so on. When the electric load in the last deuterotome of the series equals 1 eV, the series is then completed. In addition, the entire structure becomes neutral and goes through a volumetric reduction, dropping from the initial volume to reach the volume of a light hydrogen atom. The compaction of the protons and neutrons–which form the plates and the dielectrics of the atomic capacitors–gets greater and greater as each noble gas is formed in the following series with greater deuterotomic numbers. The electric load reduction of deuterons added to the structure of the previous deuterotome takes place similar to that of a liquid inside communicating vessels until a minimal possible value can be reached before it equals 1eV.

The table of Periodic Classification of Elements shows that in a given series ranging from a post-noble gas deuterotome to the corresponding halogen, the potential difference is lower and lower the higher the deuterotomic number. The electric load of valence of one or more protons, which forms the chemical pole of an element, is located on its external face.

The valences or reactive electric load of the protons in the chemical pole are available or free as far as the fourth deuterium following a noble gas–which makes up the genetic pole. They are all very high (or positive) and distribute electric load when in contact with free valences (receptors) with lower load rates (or negative). Deuterotomes of identical electric loads can combine with another if activated by heat like sulfur (tetravalent)–that combines with oxygen (bivalent)–both with an electromotive force (emf) in their valences equal to 1.33 eV when an oxygen current flows through warmed sulfur, as per the equation below:

$$S + O_2 \xrightarrow{\Delta} SO_2$$

Deuterotomes with low electric load generally yield oxides very easily, such as chlorine oxide III (Cl_2O_3), magnesium oxide (MgO), ferrous oxide (FeO) etc. These chemical combinations can be schematically represented as follows:

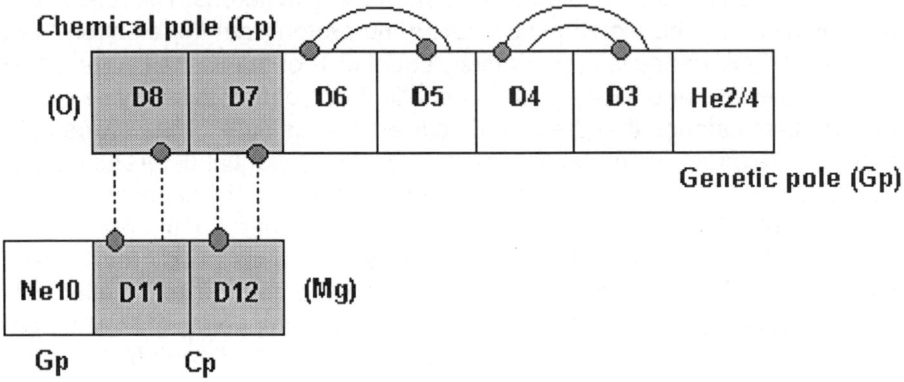

Fig. 39. Illustration of oxygen deuterotome (top) with its chemical pole formed by deuterons D8 and D7 and the genetic pole by helium-4 noble gas (He2/4). Deuterotome of magnesium is shown (below) with the chemical pole composed of deuterons D11 and D12 and the genetic pole for neon deuterotome (Ne10). The free valence of D8 and D7 deuterium atoms combines with D11 and D12. Deuterons D6/D5 and D4/D3 are connected on series-equivalent bindings.

While there are no other deuterons lying ahead, the deuterons located as far as the fourth post-noble gas unit do not have their valences blocked, that is, the reciprocal bindings fail to occur between the pair of series-equivalent connected deuterons. The next figure shows tri and tetravalent deuterotomes (the two on top) and one trivalent deuterotome, but with five post-noble gas deuterons.

The deuterons of sodium (Na11), potassium (K19), lithium (Li3) and rubidium (Rb37) will be always monovalent, for there is no deuterium with a valence available before or after the only post-noble gas deuterium that forms them. Likewise, calcium (Ca20), beryl (Be4), magnesium (Mg12) and barium (Ba56) are all bivalent and will always have two valences available and are incapable of being altered, neutralized by series-equivalent binding or having one free valence. Phosphorus, with five post-noble gas deuterons has one, three and five free valences–all with low electric load (negative, as defined by the current theory). When the first four deuterons are blocked by series-equivalent bindings, only the fifth or distal deuterium remains (D15) with an electric load of 1.66 eV to function as a free valence (monovalent).

The same can happen to nitrogen (N7) deuterotome, which can also have one, three and five free valences, with an electromotive force identical to that of phosphorus. Under thermal and baric variations, the series-equivalent bindings between the pairs of deuterons can be undone, increasing the number of free valences. Nitrogen and phosphorus can be undone, increasing thus the number of free valences. Nitrogen and phosphorus can never be heptavalent, for both of them have only five post-noble gas deuterons. The same thing happens to both vanadium (V23) and niobium (Nb41), which have the same number of post-noble gas deuterons, the last two are only tri and pentavalent and do not function as monovalent.

As a result of the physical phenomenon of volume-reduction that they are subject to, noble gases, upon becoming electrically neutral, constitute a capacitive unit with a high capacitance rate, depending on their number of protons (or plates). The distance between the plates of each unit is reduced and the value of variable **d** in the **C** capacitance equation is reduced, being equal to one for helium (He2), krypton (Kr36) and xenon (Xe54) noble gases and two for neon (Ne10) and argon (Ar18). These deuterotomes go through a volume-reduction due to the series-equivalent bindings from which they are formed. From xenon (Xe54) onwards, the bindings among the deuterons that form the deuterotome are parallel-equivalent type, which continue as far as radon (Rn86). At that stage, a different phenomenon occurs. The physical phenomenon of volume-reduction does not happen and the 32 electrons in the series of the P-shell agglomerate close to the genetic pole of the deuterotome their formed and make up what was called "electronic furnace." The electrons of francium (Fr87), radium (Ra88), actinium (Ac89), thorium (Th90), protactinium (Pa91) and uranium (U92) elements are not part of the "electronic furnace"; they only participate in the chemical pole of these elements. The post-radon deuterons can be parallel equivalent connected or not.

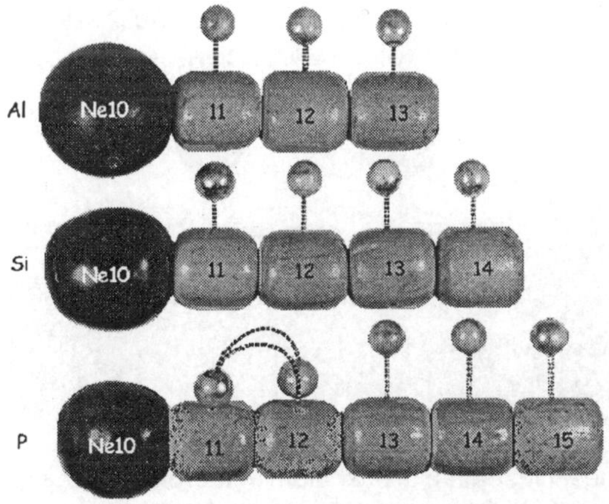

Fig. 40. Linear arrangement of deuterotomes from top to bottom is as follows: aluminum (Al13), silicium (Si14) and phosphorus (P15), all with the genetic pole formed by neon noble gas (Ne10). Note the three free valences of aluminum with an electromotive force (emf) equal to 2.66

eV in each deuterium. There are four free valences of silicium with 2.0 eV emf in each deuterium. In phosphorus deuterotome, D11 and D12 deuterons are reciprocally neutralized by series-equivalent bindings. Five is the maximum number of valences of phosphorus and their release can be obtained by thermal and/or baric actions; by blockade of the free valences D13 and D14 deuterons of phosphorus it becomes monovalent. Whatever the valence number of phosphorus (1, 3 and 5), the emf will be always the same, that is, 1.66 eV.

Because the composition of transmercury elements (Hg80) maintains a specific phylogenetic code for each radioactive deuterotome of radium (Ra88), actinium (Ac89), uranium-235 (U92/235) and thorium (Th90) families in the gap that goes from thallium (Tl81) to uranium (U92), it can be called "alpha appendix", for its composition comprises alpha-units with a variable number of neutrons, such as α2/4, α2/6, α2/8 and α2/12 with a specific distribution for each element in the radioactive families. In other words, the "locus" of each alpha-unit is specific for each transmercury element that makes up the four radioactive families.

Uranium can function with two, four or six valences. By combining with oxygen, it forms the following oxides: UO_2 (uranyl), U_2O_5 and U_3O_8, encountered in the pitchblende which is a green powder when purified; uranium trioxide (UO_3), a yellow solid that functions as amphoteric; uranium tetraoxide (UO_4), in whose molecule two oxygen deuterons are likely connected to each other.

Throughout the radioactive degradation process, the presence of oxygen or halogens linked to the valences of either uranium or radium, for instance–thus forming radium bromide (RaBr2)–in no way alters the physical phenomenon of radioactivity. The element linked to the alpha particle that is emitted breaks loose easily because its structure-features that of a noble gas, with a chemical affinity nearly null. The alpha particle emitted captures two electrons that can result from the emissions of beta particles (β) and constitute helium-4 (He2/4) deuterotome, which is why this gas can be encountered inside the ore containing both uranium and radium.

Scandium deuterotome (Sc21) functions with three valences, having three electrons in the N-shell; the distance between proton and electron is 9, corresponding to resistance of 9eΩ (electron-ohms); the electromotive force is 6 eV. The illustration of this deuterotome in the next photo shows the neutrons (dielectrics) between the protons (or plates).

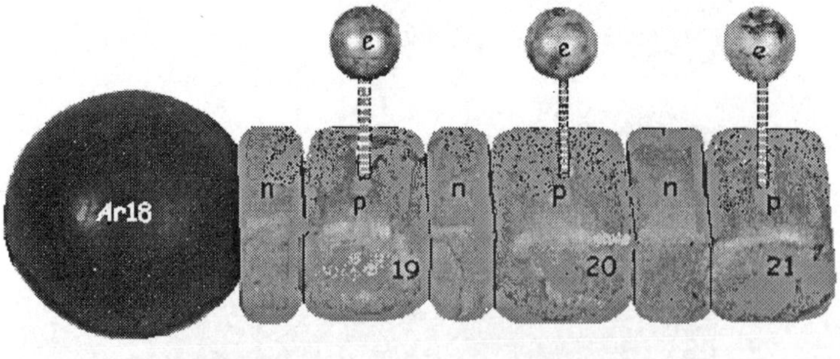

Fig. 41 – Figure is a linear illustration of scandium deuterotome (Sc21) with three free valences and an electromotive force of 6 eV. The figure shows the genetic pole constituted by argon noble gas (Ar18). There is one neutron (dielectric) between the noble gas and deuterium D19, one between D19/D20 and another one between D20/D21, acting as dielectrics. The electrons are represented over the respective protons and belong to the N-shell.

CHAPTER 7

Chemical valences. Electromotive force. Molecules.

7.1 – Electric load of valences. Chemical reactions

The distance between the proton and the orbital electron broadens as each new electron shell is formed and gets wider after each noble gas. However, as every new alpha particle is added to the previous structure of a deuterotome, the distance between the proton and the electron is successively shortened to the half, the fourth, sixth and eighth parts, when the series of the L-shell ceases, for example. Accordingly, the electric load initially of 8 eV for the lithium-6 deuterotome (Li3/6) becomes 4 eV for berillium-8 (Be4/8), 2.0 eV for carbon-12 (C6/12), 1.33 eV for oxigen-16 (O8/16) and 1 eV for neon-20 (Ne10/20). When the neon deuterotome is formed, the electric load of the series becomes 1 eV, and is electrically neutral, since there is no electric load lower than that of one proton or one electron, except in the case of quantum. Thus, as each noble gas is formed, the total electric load of the series to which it belongs turns neutral, that is, there will be no valence or incoming or outgoing electric load for the electrochemical combination to occur among the elements.

As regards deuterotomes formed through the synthesis mechanism type "neutron capture, emission of beta particle", the electromotive force (emf) yielded by the genetic pole constituted by helium-4 deuterotome, which became a deuterotomic capacitor and accumulated in lithium (Li3/6) is 8 eV. As for other deuterotome synthesized via the same mechanism–boron (B5/10), nitrogen (N7/14) and fluorine (F9/19)–the emf will be 2.66 eV, 1.66 eV and 1.14 eV, respectively, representing the third, the fifth and the seventh parts of the emf of the initial deuterotome in the series, that is, lithium (Li3/6) in the example given. The increment of an alpha particle to the structure of the deuterotome previously formed reduces the electromotive force, which is located on the external face of the proton, since the electrons occupy an internal position (intra-nuclear) in the structures of long deuterotomes, which assume a spatial arrangement in a ring-like and elliptical shape, for their composition has a great number of deuterons in their composition, as is the case of bromine (Br35) and mercury (Hg80), for example.

The table below shows:

1) The acronym of the deuterotome and its deuterotomic number;

2) The emf value for the atom of deuterium (hydrogen) and for every individual deuterotome expressed in eV,

3) The number of valences that the deuterotome can use in chemical reactions,

4) – The states of oxidation of actinides and lanthanides which bear a close resemblance to each other.

Therefore, the deuterotomes that range from lanthanum (La57) to lutetium (Lu71) and from actinium (Ac89) to lawrencium (Lr103) have all valence three (3+); deuterotomes that range from cerium (Ce58) to terbium (Tb65) and from thorium (Th90) to berkelium (Bk97) have valence four (4+); from praseodymium (Pr59) to europium (Eu63) and from protactinium (Pa91) to americium (Am95) have valence five (5+). Uranium (U92), neptunium (Np93), plutonium (Pu94) and americium (Am95) actinides have valence six, as do neodymium (Nd60), promethium (Pm61), samarium (Sm62), and europium (Eu63) lanthanides. Pm61, Sm62, Np93 and Pu94 have valence seven. Nevertheless, not all the above-mentioned valences can be frequently found in molecules or ions.

There are deuterotomes with stable valences under any condition of temperature and pressure, such as those of lithium (Li3), potassium (K19), aluminum (Al13) etc. and others with unstable or mobile valences, whose bindings are series-equivalent but capable of becoming free valences under thermal and/or baric actions, as is the case of chlorine (Cl17), phosphorus (P15), sulfur (S16), nitrogen (N7) etc.

The excess of one of the components in a reaction can fuel the valence rise in a deuterotome. For example: the excess of chlorine (Cl17) in the presence of phosphorus trichloride (PCl_3) transforms it into phosphorus pentachloride (PC_5), the equation of which can be shown as follows:

$$\mathbf{PCl_3 + Cl_2\ (in\ excess) \longrightarrow PCl_5}$$

In PCl_3 phosphorus is trivalent and in PCl5 pentavalent. The electromotive force of phosphorus is 1.66 eV for each one of the valences and that of chlorine–monovalent in that molecule–is 1.14 eV.

Hydrogen sulfide (H2S) is formed in a small volume when hydrogen current (H1/1, monovalent, emf = 2eV) goes through sulfur (S16, bivalent, emf = 1.33eV) heated up to reach the boiling point. The equation of the chemical reaction is as follows:

$$H_2 + S \underset{}{\overset{\Delta}{\rightleftarrows}} H_2S$$

Sulfur dioxide (SO_2) is dissociated under the temperature of 1,200°C, forming sulfur trioxide or sulfuric oxide (SO_3). Deuterons D11 and D12 of the sulfur deuterotome open its valences when the series-equivalent bindings are undone. Therefore, the four reactive valences that existed in sulfurous oxide (SO_2) change into six in SO_3. The equation of this reaction is as follows:

$$3\,SO_2 \xrightarrow{\Delta} 2\,SO_3 + S$$

The emf value both in sulfur and oxygen is 1.33 eV.

D1 2eV 1							He2 1eV 0										
Li3 8eV 1	Be4 4eV 2	B5 2.66eV 3	C6 2eV 4	N7 1.66eV 1/3/5	O8 1.33eV 2	F9 1.14eV 1	Ne10 1eV 0										
Na11 8eV 1	Mg12 4eV 2	Al13 2.66eV 3	Si14 2eV 4	P15 1.66eV 1/3/5	S16 1.33eV 2/4/6	Cl17 1.14eV 1/3/5/7	Ar18 1eV 0										
K19 18eV 1	Ca20 9eV 2	Sc21 6eV 3	Ti22 4.5eV 4	V23 3.6eV 3/5	Cr24 3eV 1/2/4/ 6/7	Mn25 2.57eV 1/2/4/ 6/7	Fe26 2.25eV 2/3	Co27 2eV 2/3	Ni28 1.8eV 2/3	Cu29 1.63eV 1/2	Zn30 1.5eV 2	Ga31 1.38eV 3	Ge32 1.28eV 4	As33 1.20eV 3/5	Se34 1.12eV 2/4/6	Br35 1.05eV 1/3/5/7	Kr36 1eV 0
Rb37 18eV 1	Sr38 9eV 2	Y39 6eV 3	Zr40 4.5eV 4	Nb41 3.6eV 3/5	Mo42 3eV 3/4/6	Tc43 2.57eV 4/5	Ru44 2.25eV 1	Rh45 2eV 3	Pd46 1.8eV 2/4	Ag47 1.63eV 1	Cd48 1.5eV 2	In49 1.38eV 3	Sn50 1.28e3V 2/4/6	Sb51 1.20eV 3/5	Te52 1.12eV 2/4/6	I53 1.05eV 1/3/5/7	Xe54 1eV 0
Cs55 32eV 1	Ba56 16eV 2	La57 10.6eV 3	Ce58 8eV 3/4	Pr59 6.4eV 3/4/5	Nd60 5.3eV 3/4/5/6	Pm61 4.5eV 3 at 7	Sm62 4eV 3 at 7	Eu63 3.55eV 2 at 6	Gd64 3.26eV 3/4	Tb65 2.90eV 2/3	Dy66 2.66eV 2/3	Ha67 2.46eV 2/3	Er68 2.28eV 2/3	Tm69 2.13eV 2/3	Yb70 2.0eV 2/3	Lu71 1.88eV 3	Hf72 1.77eV 4
Fr87 32eV 1	Ra88 16eV 2	Ac89 10.6eV 3	Th90 8eV 3/4	Pa91 6.4eV 3/4/5	U92 5.3eV 3/4/5/6	Np93 4.5eV 3 at 7	Pu94 4eV 3 at 7	Am95 3.55eV 2 at 6	Cm96 3.26eV 3/4	Bk97 2.90eV 2/3	Cf98 2.66eV 2/3	Es99 2.46eV 2/3	Fm100 2.28eV 2/3	Md101 2.13eV 2/3	No102 2.0eV 2/3	Lr103 1.88eV 3	Dh104 1.77eV 4
			→	Ta73 1.68eV 5	W74 1.60eV 6	Re75 1.52eV 1	Os76 1.45eV 2/3/4	Ir77 1.39eV 3	Pt78 1.33eV 4	Au79 1.28 1/3	Hg80 1.23eV 1/2	Tl81 1.18eV 1/3	Pb82 1.14eV 1/2/4	Bi83 1.10eV 3	Po84 1.06eV 6	At85 1.03eV 1	Rn86 1eV 0
			→	Ji105 1.68eV 5*	Rf106 1.60eV 6*	Bh107 1.52eV 1*	Hn108 1.45eV 2/3/4*	Mt109 1.39eV 3*	xx110 1.33eV 4*	xx111 1.28eV 1/3*	xx112 1.23eV 1/2*	xx113 1.18eV 1/3*	xx114 1.14eV 1/2/4*	xx115 1.10eV 3	xx116 1.06eV 6*	xx117 1.03eV 1	xx118 1eV 0*

Fig. 42 – Table of deuterotomes including the heavy atom of hydrogen (deuterium), starting from the first post-noble gas deuterium of the series in the respective electronic shells (L, M, N, O, P, Q), indicating: 1) the acronym of the deuterotome with its respective protons number; 2) the value of the electric load which the onset of the series expressed in electron-volt (eV); 3) the valences number that the deuterotome often produces in chemical reactions. Not all the valences of transuranium elements are referred to. The elements ranging from 110 to 118 were not yet synthesized but the electric loads of their deuterons will most likely be equal to those found in the elements belonging to the P-shell series. Their respective valences are also shown with an asterisk (*).

7.2 – Neutralization of the electric load of noble gases and the physical phenomenon of volume-reduction.

Whenever a series of deuterons constituted by numbers 2 (K-shell), 8 (L-shell and M-shell), and 18 (N-shell and O-shell) is completed, the series becomes electrically neutral and the volume in the whole series is reduced, including the preceding one. The physical phenomenon of volume-reduction occurs as a result of the neutralization undergone by the deuterons that compose the electronic series, no matter if they have 2, 8 or 18 electrons. In order for Avogadro's number (6.06×10^{23} molecules-gram/cm^3) to be valid for all noble gases, whichever their mass number, the volumetric reduction of the deuterons that make it up must take place, the more intense or percentually higher their mass number. A theoretical comparison can be made between the volumetric reduction of noble gases and the matter captured by "black holes" in space; both reduce the volume of deuterotomes and increase their specific weight enormously (density). The density of each noble gas is similar to its mass number expressed in grams/cm^3. Therefore, one xenon (Xe54) deuterotome occupies the volume of one hydrogen atom, but its density is 131.3 times higher. The same comparison can be made with respect to argon (Ar18), the density of which is 19.97 and krypton whose density is 83.7 g/cm^3.

The volumetric reduction of the protons and neutrons that participate in the composition of deuterons, which make up the noble gases, is not compression from the physical point of view. The physical phenomenon of volume-reduction is a specific physical property resulting from the fact that deuterons are really capacitors, series-equivalent connected. Parallel equivalent bindings in trans-xenon deuterotomes do not trigger this phenomenon. The volume-reduction phenomenon is more commonly and entirely seen in noble gases and should occur as happen to other gaseous elements at room temperature. Such is the case of chlorine, bromine, nitrogen etc, in their molecular states.

It can be said that "the volume-reduction percentage in a certain noble gas is directly proportional to the atomic mass of hydrogen (H1/1) taken as pattern and inversely proportional to the gas mass number". Hence, the volume-reduction percentage in each noble gas is as follows: 75% for He2, 95% for Ne10, 97.5% for Ar18, 97.62% for Kr36, 98.14% for Xe54. The volume-reduction percentage varies among isotopes, albeit small.

7.3 – Electric load of protons of each deuterotome

It can be also said that "the reduction in the electric load of the proton that integrates a certain deuterotome, results from the fact that the distance between the proton and its orbital electron gets shorter and is directly proportional to the electronic load in the previous proton and inversely proportional to the number of protons making up the series between an earth alkaline and a noble gas". Such being the case, in the K-shell, we have: 2 eV for D1 and 1 eV for He2. In the L-shell we have: 8 eV for Li3; 4 eV for Be4; 2.66 eV for B5; 2.0 eV for C6; 1.66 eV for N7; 1.33 eV for 08; 1.14 eV for F9 and 1.0 eV for Ne10 (which is neutral). These values correspond to the electromotive force of protons.

The low density of a metal is not associated with the rise in volume of the deuterotome, but rather with the crystallization process, resembling of porous net. Lithium (Li3) is the metal of smaller density, being a resemblance of a porous net. Lithium (Li3) is the metal with the lowest density, which equals 0.59 g/cm³. Potassium (K19) whose mass number is higher than that of sodium (Na11)–density of 0.87 g/cm³–while the latter has a density of 0.97 g/cm³. However, the high density of noble gases is associated with their volumetric reduction because of the neutralization undergone by their deuterons through series-equivalent bindings, with a load of 1 eV. The specific weight (density) of a noble gas is dependent on its mass number and, consequently, varies in keeping with the mass number in each isotope. All the noble gases are provided with isotopes, including helium-4 (He2/4), whose isotope has a mass number of 3 (He2/3), existing, however, in a minimal percentage (0,000138%).

The electrons located on the orbital zones in an electronic series with 2, 8 and 18 electrons, neutralize electric loads of pairs of deuterons until they become neutral at the end of the series. Nevertheless, from the highest electric load encountered in the first post-noble gas deuterium up to the lower one which is verified in halogens (F, Cl, Br, I)– the last deuterons in each series– there is a progressive reduction in the electric load whose value is a fraction between the initial electric load in the first deuterium and the successive numbers in each deuterium of the series, expressed in electron-volt (eV). At every two deuterium atoms coupled to the deuterotome structure through alpha particle capture, a reciprocal neutralization is set up, that is, the orbital electron of a proton neutralizes the proton load with which the pair is formed. The electrochemical phenomenon is established at the very time the coupling of the sixth post-noble gas occurs. This phenomenon fails to happen to the fast pair; otherwise the deuterotome would become a null electric load, similar to a noble gas. Accordingly, the fifth and sixth deuterium atoms that make up a series of eight deuterons have free electric loads and, as a consequence, two free valences, as is the case of oxygen (O8) and sulfur (S16), for example. As soon as the seventh deuterium atom is aligned with the deuterotome structure, the three former pairs of atoms are neutralized, leaving a valence free for the seventh deuterium, which has the lowest electric load in the L-shell and M-shell series. Baric and thermal variations can undo pairs of neutralization, providing a progressive rise in the number of valences to the deuterotome. In the case of chlorine, for example, when the valence is single, only the seventh deuterium is electrically active; when the bindings between the fifth and sixth deuterons (D15/D16) get undone, chlorine deuterotome starts to have three free valences via the same mechanism; when the valences of first and second deuterons (D11/D12) are free through the same mechanism, chlorine then has seven valences; with the bindings undone between deuterons D13/D14, chlorine has five valences. However, the valences only get free in the sequence from D15/D16 to D13/D14 and from the latter, to D11/D12.

The valences of iron (Fe26) deuterotome, for example, have an electromotive force of 2.25 eV. Deuterons D25/D26 have free valences (iron divalent). In the pair of deuterons D23/D24, the D24 valence can be released and iron becomes trivalent. In this case, the pair of deuterons D23/D24 will only have D24 with free valence. The valences blocked by neutralization can become free as a reason of thermal and baric variations and the excess of a reagent over the other, besides other factors such as chemical affinity, hydrogen potential (pH) of the medium etc. The illustrations below show the iron deuterotomes–divalent and trivalent.

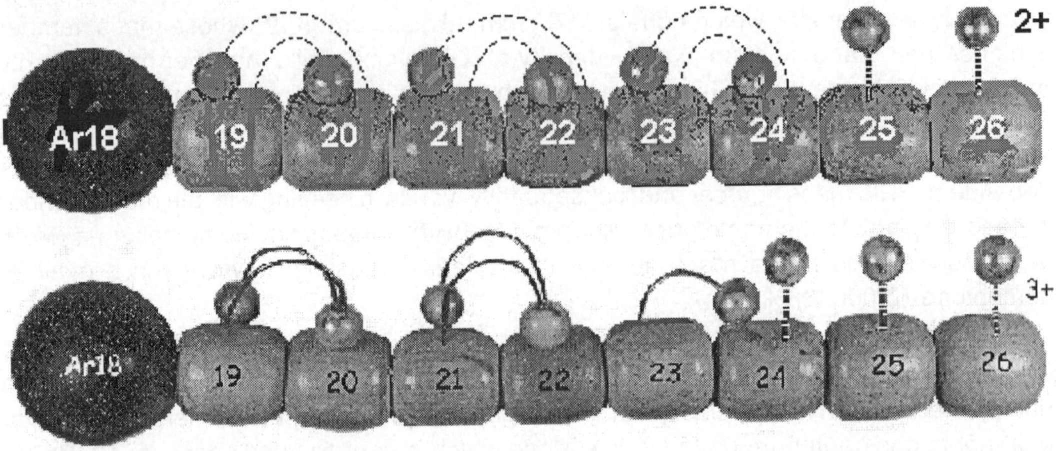

Fig 43 - **Above**: linear arrangement of divalent iron deuterotome (Fe26) with the genetic pole constituted by argon noble gas (Ar18) and its post-noble gas structure by deuterons numbered from D19 to D26, from left to right; the chemical pole constituted by deuterons D25 and D26, both with free valences (divalent). The pairs of deuterons D19/D20, D21/D22 and D23/D24 are reciprocally neutralized by series-equivalent bindings. **Below**: linear arrangement of trivalent iron with the electric loads D24, D25 and D26 activated. The electromotive force of each valence is equal to 2.25 eV.

7.4 – Molecular composition of chemical substances. Neutral and ionized molecules (in hydrolysis)

Polar substances are the ones capable of:

1) – Suffering ionization, such as elements Na+, K+, Cl–, Ca++ etc.,
2) – Reacting chemically by way of their valences,
3) – Entering ebullition to high temperatures, and
4) – Being destitute of isomeric.

Non-polar substances fail to have these properties.

The following figure is a simplified illustration of a hydrogen atom and a chlorine deuterotome, showing that they can combine chemically to each other, depending on their chemical affinities, valences available and electromotive force of the deuterons. For the purpose of illustration, in the following photos, the molecules of hydrofluoric acid (H2F2), hydrochloric acid (HCl), beryl oxide (BeO), calcium oxide (Ca0) and lithium hydroxide (LiOH) are represented. Nevertheless, none of these illustrations show the molecule electrically neutral; these are shown elsewhere.

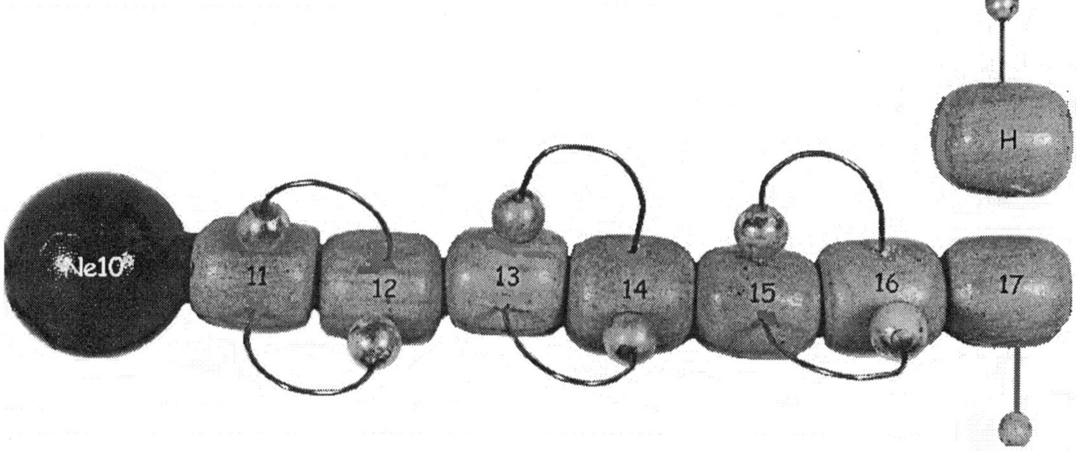

Fig. 44. **Below**: linear arrangement of the structure of chlorine deuterotome (Cl17), showing the genetic pole to the left, constituted by noble gas neon (Ne10) and followed by deuterons from D11 to D17. The deuterons are with the electric loads neutralized to their pairs from D11 to D16. In the chemical pole, the D17 has a free valence and an electromotive force of 1.14 eV. **On top**: illustration of light hydrogen atom (H1/1) with its free valence and electromotive force of 2 eV (positive, in relation to chlorine).

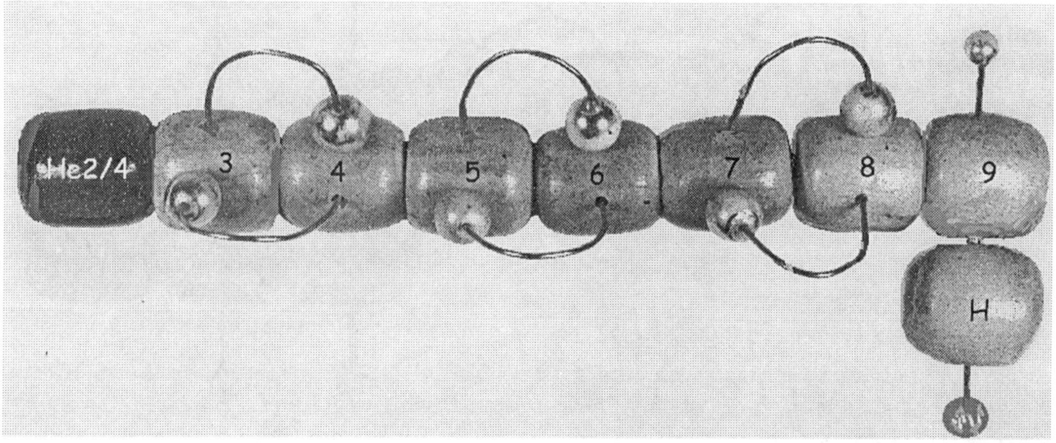

Fig. 45 – Illustration of hydrofluoric acid molecule (H2F2). Deuterium atom D9 is combined with light hydrogen atom (H1/1). The genetic pole of fluorine is noble gas helium (He2/4); the chemical pole is deuterium D9.

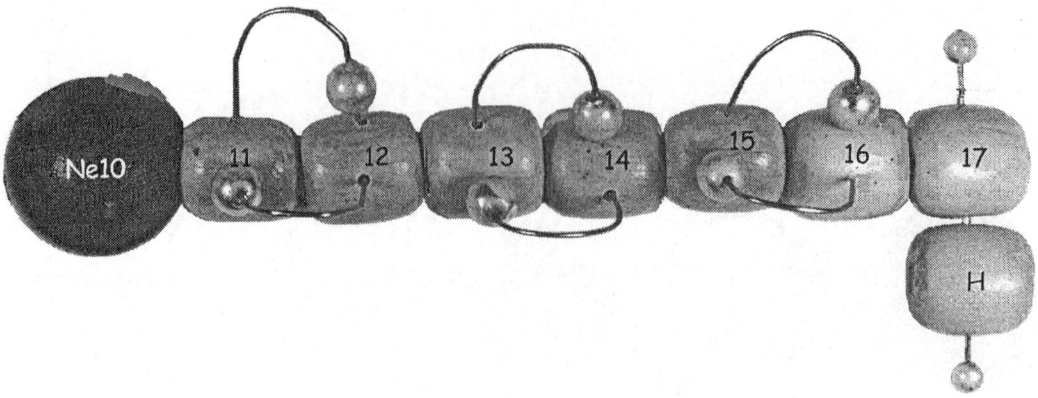

Fig. 46 – Illustration of hydrochloric acid molecule (HCl). Deuterium D17 is combined with light hydrogen atom (H1/1). The genetic pole of chlorine is constituted by neon noble gas (Ne10) whose volume is twice that of helium (He2/4). The chemical pole of chlorine is deuterium D17, when it presents only one valence (monovalent); deuterons D15, D16 and D17 represent the chemical pole when chlorine functions as trivalent. Chlorine pentavalent and heptavalent have deuterons D13/D14 and D11/D12 free, respectively, besides the other aforesaid deuterons.

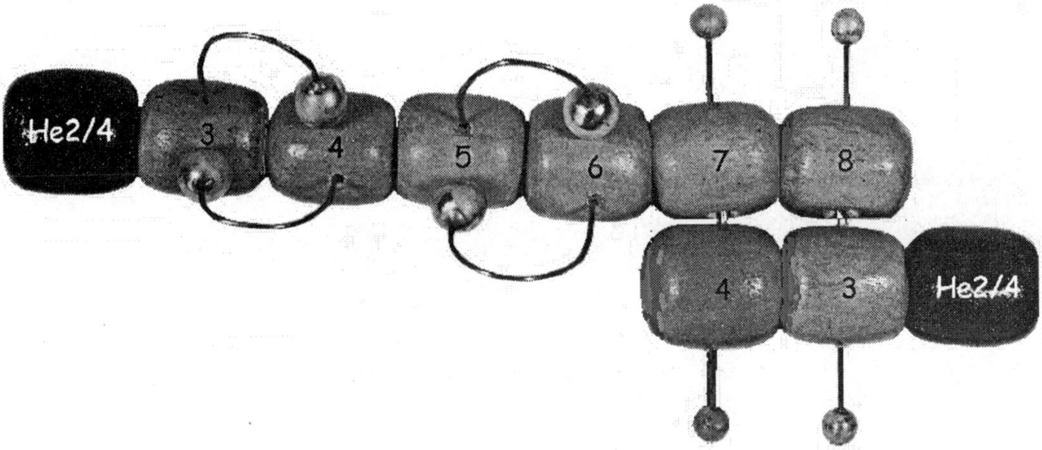

Fig 47 – Illustration of beryl oxide (Be0) molecule. The two deuterotomes have their genetic poles constituted by helium-4 noble gas (He2/4). Deuterons from D3 to D8 compose the structure of the deuterotome of oxygen (upper position of the photo); deuterons D3 and D4 compose that of beryl (lower position of the photo). Both beryl and oxygen are divalent; beryl has an electromotive force of 4 eV and oxygen 1.33 eV.

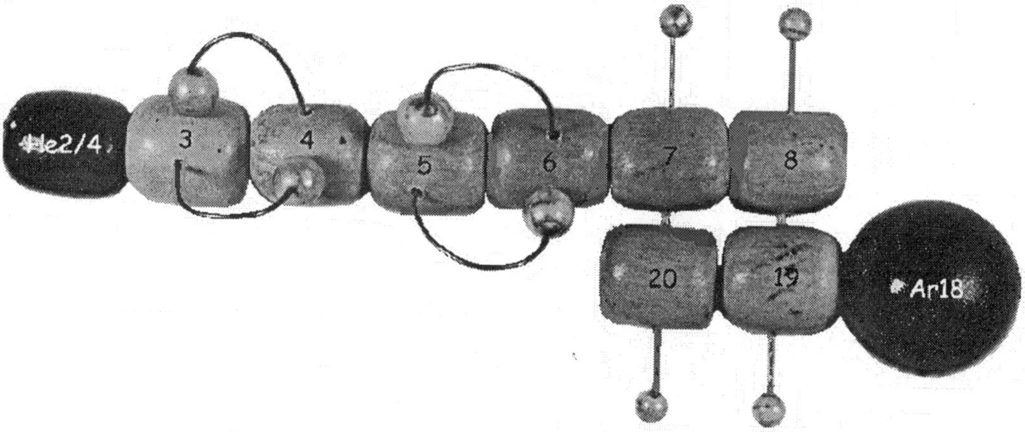

Fig.- 47-a - Illustration of calcium oxide molecule (Ca0). The genetic pole of oxygen is composed by helium-4 (He2/4) noble gas (upper position) and calcium by argon (Ar18) noble gas; its volume is twice greater than that of helium. Deuterons from D3 to D8 compose the structure of oxygen and deuterons D19 and D20 that of calcium. Oxygen and calcium have, in their chemical pole, two available valences each. The electromotive force of the deuterons of calcium is 9 eV, and that of oxygen 1.33 eV.

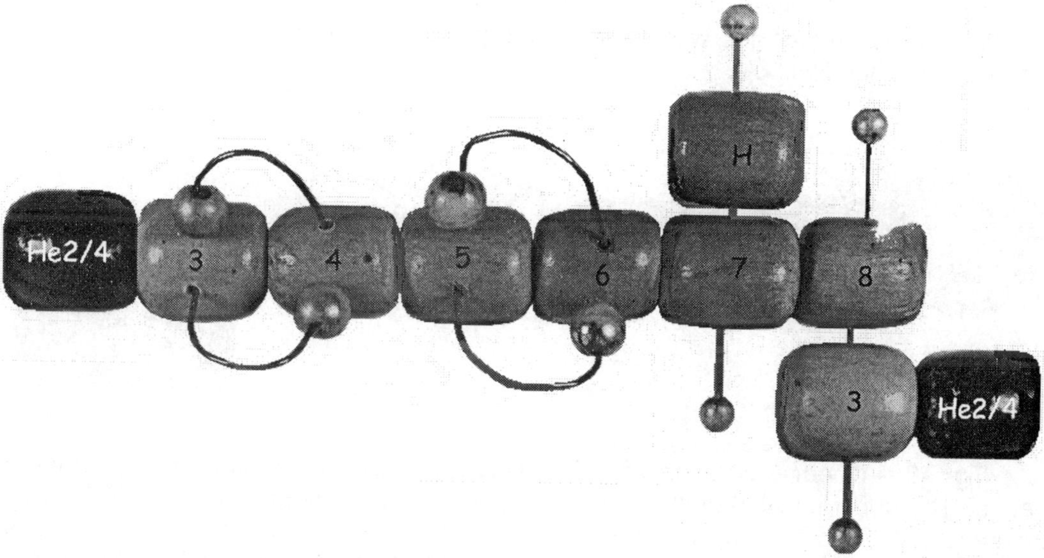

Fig. 48 – Illustration of lithium hydroxide molecule (LiOH). Oxygen and lithium deuterotomes have their genetic poles composed by helium-4 (He2/4) noble gas. Deuterons from D3 to D8 compose the structure of oxygen. The deuterium atom that composes the structure of lithium is single (D3). The two valences of oxygen are linked to hydrogen (H) and to lithium (Li), respectively.

7.5 – Molecules. Examples of chemical combinations. Electrical neutralization of molecules

The atom of hydrogen neither has a have genetic pole nor a chemical one. It is electrically neutral in its molecular form (H_2) and chemically active in its atomic or ionic form. The two atoms of hydrogen that compose the molecule of hydrogen are combined chemically to each other, different from a synthesis binding. This fact must be highlighted because, in the case of heavy hydrogen (deuterium), the molecular form is also diatomic (D2) and averages two protons and two neutrons or two protons and four nucleons, that is, the same composition as helium-4 (He2/4) deuterotome. However, between two deuterium atoms that compose the deuterium molecule (D2) there is a **chemical combination**, while between the two deuterons that compose helium-4 deuterotome (He2/4), there is a **synthesis binding**. In the deuterium molecule there is no contact potential between proton and neutron, while in helium deuterotome it is available.

The chemical or electrochemical reactions are combinations that happen among the deuterons located in the chemical pole of a deuterotome with the deuterons of the chemical pole of another. A few factors such as the number of valences free from the deuterotomes existing in the same medium so that they can react to each other, an electromotive force of each free valence and the potential difference among the free valences for reaction will provide: 1) the rate by which the deuterotomes will react to each other; 2) the amount of reagents that will react to each other; 3) the force of electrochemical combinations among the reagents.

The following examples show illustrations of lithium chloride (LiCl) and calcium chloride (CaCl2) molecules with their electric loads neutralized in the deuterons that make up the chemical pole.

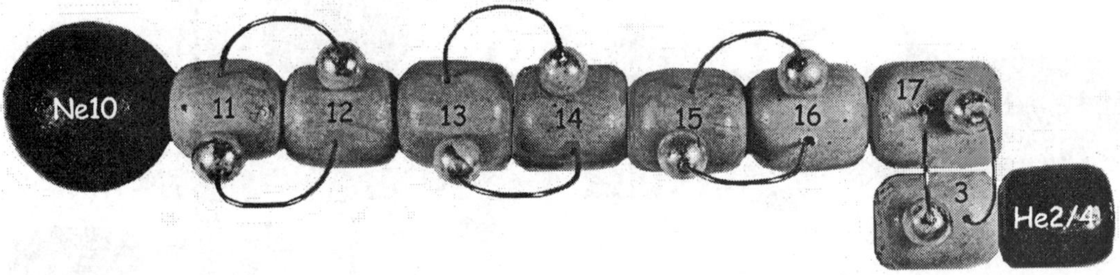

Fig. 49 – Linear illustration of lithium chloride (LiCl); this molecule is electrically neutral. The upper position of the molecule represents chlorine deuterotome (Cl17) with the genetic pole (to the left) constituted by neon (Ne10) noble gas; the post-noble gas structure is formed by deuterons that range from D11 to D17. Deuterium D17 is the chemical pole of chlorine and is combined with deuterium D3 of lithium (Li3) that has the genetic pole composed by helium-4 (He2/4) noble gas. The electromotive force of lithium is 8 eV and that of chlorine, 1.14 eV. Deuterons D17 and D3 are neutralized to each other, consequently, the molecule is electrically neutral.

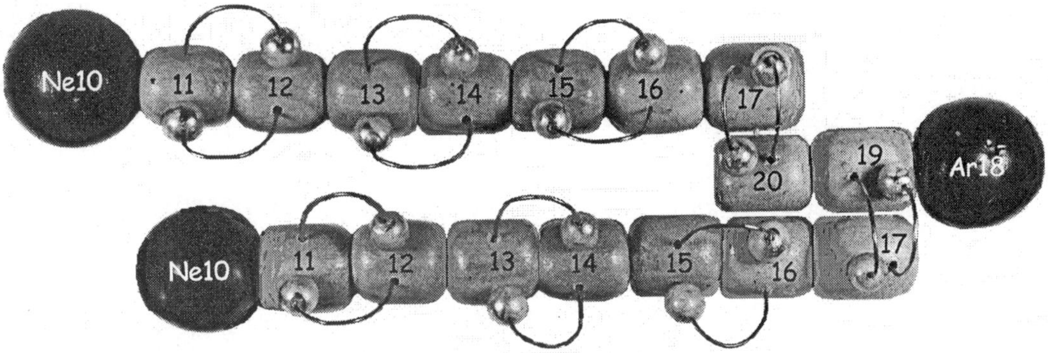

Fig. - 50 – Linear arrangement of calcium chloride ($CaCl_2$) molecule: it is electrically neutral. These two structures, upper and lower, are the deuterotomes of chlorine (Cl17) with the genetic pole constituted by neon noble gas (Ne10). Deuterotome of calcium, between the two of chlorine, has the genetic pole constituted by argon noble gas (Ar18). The chemical pole of calcium is constituted by deuterons D19 and D20 and that of chlorine by deuterium D17. The electromotive force of calcium is 9 eV and that of chlorine 1.14 eV. Deuterium D17 is combined with D19 of calcium and another D17 with D20, turning the molecule electrically neutral.

CHAPTER 8

Radioactivity, atomic projectiles and radioactive emissions

8.1 – Sources of radiations

Radiations that can be found throughout the process of nuclear fusion, nuclear fission, radioactive degradation, production of cathode rays and others, are made up of particles with electric load and particles and radiations without it.

Radiations with electric load are the fast electrons (beta-negative particles, positive (e+) and negative (e^-) energetic electrons) and heavy particles (alpha, protons and fission products and from other nuclear reactions).

Radiations without electric load are: 1) photon radiations (X rays, emitted as a result of the rearrangement of electrons shells and gamma rays, originating from transitions inside their own nucleus); and 2) neutrons particles (slow and fast).

Alpha (α2/4) and beta (β) particles and neutrons are the three most important radiations for the synthesis of deuterotomes and the process of radioactive degradation

The radiation energy is emitted or absorbed in small flake or quantum; a quantum of energy is a photon. A photon or a quantum is the smallest amount of energy emitted or absorbed when it is displaced from a higher energy level to a lower one or vice-versa. Each photon of radiation of frequency **v** has an energy **E** in accordance with the following equation: **E = h.v**, where **E** is the energy of photon, h is Planck's constant (h = 6.62618 x 10^{-34} J.s) and **v** is the radiation frequency. Accordingly, the higher the frequency of a photon the higher its energy. The photon of blue light (λ = 450 nm) has far more energy than the photon of red light (λ =700 nm) because of its greater frequency.

8.2 – Source of fast electrons

The common source of fast electrons in radiation units is a radioisotope, whose beta decay occurs due to the emission of beta particles (β^-). The process is schematically illustrated as per the following formula:

$$X_Z^A \text{ ------------------------> } Y_{Z+1}^A + \beta^- + \bar{\nu}$$

where **X** and **Y** are the initial and final deuterotomes and □ν the antineutrino. **A** is the deuterotomic mass or mass number. The antineutrino, which is antimatter, has little likelyhood of interacting with matter and is undetectable. Therefore, the only significant ionizing radiation produced by beta decay is the fast electron; in other words, beta particle

(β^-) itself. A large number of radionuclides produced by the bombardment of neutrons from stable deuterotomes are beta-active. A wide variety of beta emitters is quickly available in flow nuclear reactors. Isotopes of different half-lives can be obtained in these reactors, ranging from a few hundred years to a few days or even seconds. In many sources of beta particles (β^-), gamma rays are emitted along with that particle. However, a few radioactive isotopes can only emit beta radiation. Some examples of these can be mentioned with their respective half-lives: C6/14 (5,730 years); P15/32 (14.2 days); Tl81/204 (3.81 years); Cl17/36 (3.08 x 10^5 years). The respective equations are as follows:

C6/14————> N7/14 + β^- Tl81/204————>Pb82/204 + β^-

P15/32————> S16/32 + β^- Cl17/36————>Ar18/36 + β^-

8.3 – Sources of heavy charged particles

8.3.1 – Alpha particles emitted by natural radioactivity. Heavy deuterotomes (Z > 81) of the radioactive family of radium, actinium, uranium-235 and thorium, as well as transuranium elements are energetically unstable, unable to uphold the number of their protons (Z), and emit alpha particles (α 2/4). The likelyhood of beta decay is governed, in keeping with the theory in force, by a barrier penetration mechanism. The alpha decay process can be schematically represented as follows:

$$X_{Z}^{A} \text{ ------------------------->} Y_{Z-2}^{A-4} + \alpha\ 2/4$$

where **X** and **Y** are the initial and final deuterotomes, respectively. A and B are the mass and deuterotomic number, respectively.

Alpha particles (α2/4) come up in one or more groups of energy that are mono-energetic. A clear correlation exists between the kinetic energy of an alpha particle and the radioactive half-life of isotopes. Isotopes provided with higher energy levels are those with shorter half-life. When the radioactive isotope has an alpha particle (α2/4) kinetic energy somewhat over 6.5 MeV, its half-life can be expected to be less than a few days. Should that energy level drop below 4 MeV the likelyhood of the barrier penetration mechanism to act is slim and the isotope half-life becomes higher. When the isotope half-life is far too long, the radioactive intensity is then very low. Thorium-232 (Th90/232), for instance, has a long half-life of 1.4 x 10^{10} years and the alpha particle (α2/4) kinetic energy it emits is lower, equal to 4.012 MeV. Einstenium-253 (Es99/253) has a short half-life of 20.5 days and a high alpha particle kinetic energy of 6.632 MeV. The equation below illustrates this type of radioactivity:

Ra88/226 ——————————>Rn86/222 + α2/4

The half-life of radium (Ra88/226) isotope is 2,205 years, a million times more radioactive than uranium (U92); in the radioactive family of radium it emits an alpha particle (α2/4) producing an element of deuterotomic number and a lower mass number in accordance with the scheme X——> Y + α2/4, thus creating radon-222 (Rn86/222), the half-life of which is 5.54 days.

8.3.2 - Spontaneous fission. The fission process is the only spontaneous source of energetic heavy particles with a mass number higher than that of alpha particles, largely used to gauge and test radioactive matters. This fission process occurs in some trans-uranium isotopes with great mass number. The element more often employed is californium-252 (Cf98/252), which undergoes spontaneous fission with a half-life of 85 years; it is a source for alpha particles (α2/4) because a microgram emits 1.92 x 10^7 of alpha particles. It is submitted to 6.14 x 10^5 of spontaneous fissions/second, also emitting fast neutrons.

8.4 – Alpha particle (Kα) kinetic energy and Excess of mass energy (Q)

Alpha particle (Kα) kinetic energy emitted during the alpha decay process in the radioactive degradation of radium, actinium, uranium-235 and thorium all of which are elements of radioactive families, can be calculated in accordance with the following equation:

$$K_\alpha = \frac{A-4}{A} \cdot Q$$, where:

Kα is the kinetic energy of alpha particle (α2/4),
A is the mass number of the element, and
Q is the excess of mass energy. **Q** is calculated by the following equation:

Q = [(mXn – mXn-2 – m (α)] x C^2, where:

Xn is the mass number of the element emitting alpha particle,
Xn-2 is the mass number of the element formed after the loss of mass of alpha particle,
m (α) is the mass of the alpha particle; and
C^2 is the square of light speed.

In order to calculate the **Q** value during the radioactive degradation of thorium-230, the following equation is used:

Th90/230 ——————————— >Ra88/226 + α2/4, where:

m of **Xn**, that of Th90/230 = 230.033131u
m of **Xn-2**, that of Ra88/226 = 226.025406u
m (α), the mass of the alpha particle = 4.00150618u
C^2, the square of light speed = 931.5 MeV/u

By applying the equation in **Q**:

Q = [230.033131u – 226.025406u – 4.00150618u] x 931.5 MeV/u

Q = 0.0062239u x 931.5 MeV/u

Q = 5.96523 MeV

To calculate the kinetic energy of the alpha particle the Kα equation is used:

$$K_{\alpha} = \frac{A-4}{A} \cdot Q \text{ , where:}$$

A = mass number of element **Xn**; in this case that of Th90/230;

Q = excess of mass energy in the alpha decay process; in this example 5.96523 MeV.

Kα = 230 – 4/230 x Q

Kα = 5.6967 MeV

Thus, the kinetic energy of an alpha particle is 5.69 MeV.

According to the classic atomic theory, the barrier of a typical heavy nucleus lies between 30 and 40 MeV. Because the alpha particle has a kinetic energy ranging from 4 to 8 MeV, in no way can it abandon the nucleus and overcome an energy barrier five to seven times higher than its own energy level. For that reason, the theory acknowledges that the only possible way for the alpha particle to escape from the nucleus and beat the penetration barrier is through the "tunnel", across the barrier.

However, in light of this new theory–the deuterotomic theory–the alpha-unit situated at the very end of the chemical pole of heavy deuterotome can be removed by a rupture in the synthesis binding between the alpha-unit and the remaining deuterotome structure, which will then have its mass number decreased by four units and the number of protons by two units, thus not requiring the tunnel to overcome the penetration barriers. For this to happen only the binding energy per nucleon must be overcome, the value of which depends on the heavy deuterotome it belongs to. For the sake of comparison, one can cite that the binding energy per nucleon (Ben) in U92/235 is 7.591 MeV and in U92/238, 7.571 MeV.

Bismuth-210 (Bi83/210), which emits alpha particles during the process of radioactive degradation of the actinium family, has a total binding energy (Tbe) of 1,648 MeV and a binding energy per nucleon (Ben) 7.849 MeV, the values of which comply with the range of kinetic energy of alpha particles. Thus, the real difficulty in interpretating a radioactive phenomenon, by adopting the atomic model of Rutherford-Bohr and Thomson, with a central nucleus and orbital electrons revolving around (external electronic crown), is not the same as that found in the deuterotomic model consisting of the association of unitary atomic capacitors (protons plates and neutrons dielectrics) and orbital electrons of intradeuterotomic form (electronic crown located internally).

The Q value in beta decay: negative and positive. Overall, the negative beta decay is represented as follows: n——> p + e^- + □ν, where **p** is the proton; $\mathbf{e^-}$ the negative electron and□ν, the antineutrino. The antineutrino mass (m) equals zero. The following is an example of beta decay:

08/19 ————>F9/19 + e^- + □ν

During the emission process of beta particles (β^-) the **Q value** (excess of mass energy) is 4.82 MeV. Deuterotomes with higher mass number produce lower Q values. During the process of positive beta decay, the scheme is represented by: p——> n + e^+ + v, where **p** is the proton, **n** the neutron, **e+** the positron or anti-electron and **v** the neutrino. A reaction example can be figured out:

Al13/25——————>Mg12/25 + e^+ + v

The process of nuclear decay which opposes the emission of positron in the positive beta decay consists of electron capture, the basic equation of which is as follows: p + e^- ———> n + v, where a nuclear proton captures an electron in the orbital zone and is converted into one neutron plus one neutrino. This capture process fails to occur in free protons, but it does in protons situated inside the nucleus and in one of the two protons that make up the transitory unit of bi-hydrogen (H2/2) which is converted into neutron, thus creating the atom of deuterium (D1/2). A typical example of the said beta decay processes is shown below:

1) Tm69/170 + e^- ------------------------> Er68/170 + v

2) H1/1 + H1/1 ---------fusion---------> H2/2 ---------conversion---------->D1/2 + v + e^+

Gamma decay - After an alpha (α2/4) or a beta (β^-) decay has occurred the nucleus can be led to a state of excitation, either by the electron hitting the deuterotome wall in the chemical pole, where the alpha-appendix lies, or by a rupture in the link between the group of deuterons D92/91 with D90 that integrates the rest of the deuterotome, as is the case of uranium-238 (U92/238) or the emission of beta particles (β) via the neutron and its consequent transformation into a proton. Through the shock of electrons and the rupture in the links, the nucleus will then abandon its ground state after emitting one or more photons (in the case of beta particles emission); such photons are known as **nuclear gamma rays**. The amount of energy in each photon is exactly the same as the energy difference found between the initial nuclear state (i) and the final one (f), minus the negligible small correction for the nucleus recoil energy, and these energy levels are typically around 10 keV and few MeV. Below is a diagram of energy levels:

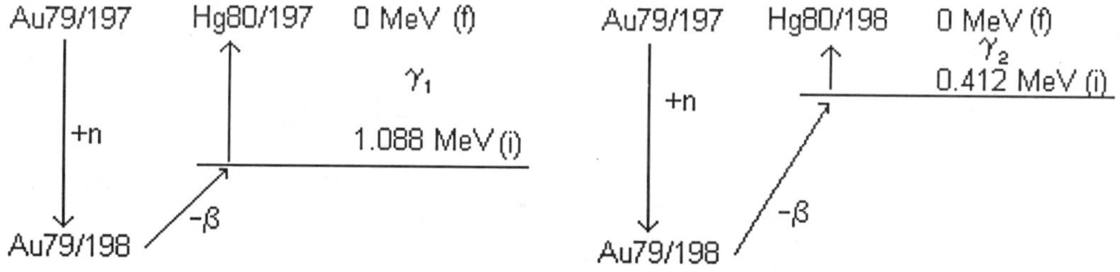

Fig. 51. In gold beta decay (Au79/198) the energy **E** of photons (nuclear gamma rays) is represented by a diagram of energy levels. Depending on the nuclear excitation state, the difference between the initial (i) and final (f) energy will determine frequency v of the photon emitted in keeping with the equation **E = h.v.** On the left picture the energy of gamma ray 1 (γ1) is 1.088 MeV; and on the right picture the energy of gamma ray 2 (γ2) is 0.412 MeV.

The alpha and beta decay processes that happen in the chemical pole of the heavy deuterotome can be schematically represented. For the sake of analysis, the heavy deuterotome can be represented in oval shape (like Crookes' tube) with its three basic structures: **A** – genetic pole, consisting of noble gas xenon (Xe54); **B** – chemical pole, formed by the "alpha-appendix" where the decay occurs; **C** – "electronic furnace", composed of 32 electrons from where intranuclear atomic projectiles (electrons-projectiles) depart, that are thrown against the proton/neutron (P/N) junctions, causing ruptures in them, emitting alpha particles (α2/4).

Fig. 52 shows a microscopic structure of the wall of uranium (U92/238) where chemical pole lies, formed by the "alpha-appendix". As the binding energy per nucleon (Ben) is different in each deuterotome, the impact of the electron-projectile against the P/N junction consequently generates a different excited nuclear state. The binding energy per nucleon decreases upon the emission of alpha particle (α2/4) and increases, far beyond the value prior to the emission of that particle after emitting the beta particle (β). Both the collision of the electron-projectile against the P/N junction–and its consequent rupture–and the emission of the electron linked to the neutron, produce an excited nuclear state, that leads to the nuclear emission of gamma ray photons, whose energy **E** will depend on the initial (i) and final (f) excited nuclear state of the deuterotome.

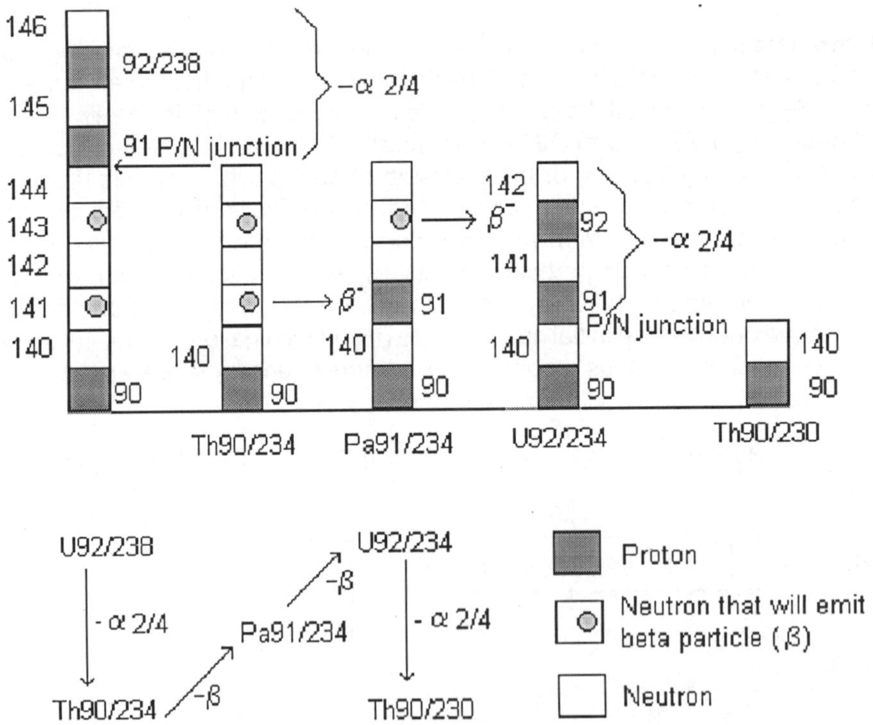

Fig. 52. Scheme of initial alpha decay of U92/238: 1) U92/238 emits alpha particles (α2/4) with the rupture of the proton/neutron (p/n) link in the position indicated by the word "junction"; reduces the deuterotomic number and deuterotomic mass and is changed into thorium-234 (Th90/234); the emission of beta particles (β) by neutron-141 and, before, by neutron-143 resulting from the excited nuclear state with initial (i) and final (f) energy that transform thorium-234 and protactinium

Pa91/234 into uranium-234 (U92/234) because neutron-141 and neutron-143 acquire a proton-like status; 2) uranium-234 emits alpha particles because it was designed to do so (alpha particles with more than two neutrons in their structure can in no way be emitted); emission of beta particles (β), whose electron originates in neutron-141 and neutron-143; as a result of alpha particles (α2/4) being emitted, protactinium-234 is transformed into uranium-234 and upon beta emission (β) the latter is changed into thorium-230 (Th90/230), thus able to continue on emitting alpha particles.

8.5 – Electrons of Crookes' tube as "atomic projectiles"

Electrons emitted by the cathode of Crookes' tube are carried forward by the deuterotomes of rarefied gases and ionized gases in its interior, making up **cathode rays**. An electric load of 30,000 volts, for example, is applied to the tube's cathode thus accelerating electrons (electrons-projectiles) in opposite direction, hitting the tube glass-wall. During the interaction of the electron-projectile with the glass-wall, electromagnetic radiations are produced during the rearrangement of electrons in their orbital shells. Manifestations occur in the target-matter owing to this interaction: 1) photoelectric absorption; 2) Compton's effect; 3) bremsstrahlung and 4) production of a pair of electrons (electron + positron).

8.6 – Effects of emission and radioactive degradation

The spiral-shaped form of the heavy radioactive deuterotome composed of the genetic and chemical poles and the internal electronic furnace, fuels several physical phenomena, true mechanisms that sustain the emission of gamma ray photons (γ), alpha particles (α2/4) and beta particles (β). Inserted in these mechanisms are the fundamentals of the two Laws of Thermodynamics, that is, the transformation of work in heat (1st Law) and heat in work (2nd Law).

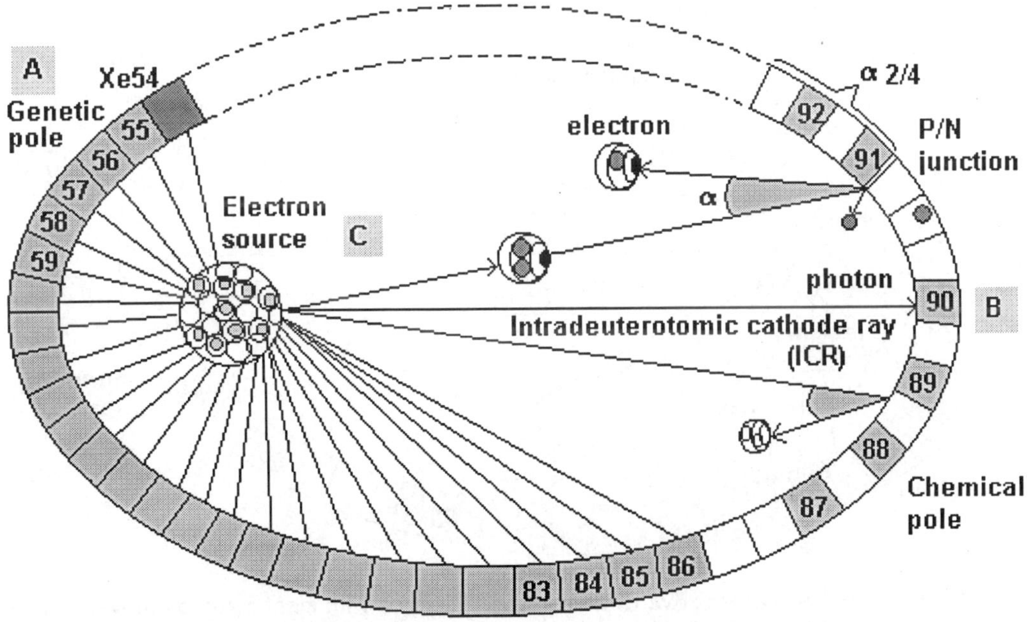

Fig. 52-a – **Heavy Radioactive Deuterotome** – Electron released from the source (electronic furnace) act as an intradeuterotomic cathode ray (ICR). Throws against the chemical pole, it causes the P/N junction rupture, releasing alpha particles (α2/4). The electron can also deflect and form an alpha angle and emit intradeuterotomic photons.

8.6.1 – Furnace electron as possible intradeuterotomic cathode ray

The 32 electrons that make up the electronic furnace of the radioactive deuterotome (U92/238 for instance) that are thrown against the P/N junction located in the chemical pole area, act as intradeuterotomic cathode rays (ICR). An electron coming from the source (electronic furnace) needs kinetic energy (Ke) equal to or higher than the binding energy per nucleon (Ben) to cause the rupture of P/N junction. Therefore, if the P/N junction of the U92/238 has a Ben of 7.571 MeV, the kinetic energy (Ke) of ICR will have the same or higher energy for it can cause the rupture of that junction. The energy necessary for the electronic furnace to emit ICR can be obtained by the transference of energy from the Xe54 deuterotomic capacitor to its electrons, conferring upon them high energy. Xenon has a capacitance value of 84.8232 eF.

8.6.2 – Work and Heat in the dynamics of the radioactive deuterotomes

The electron, whose mass **m** is 0,511323 MeV/c^2, moves in accelerated motion (g) towards the P/N junction of the chemical pole. The kinetic energy (Ec) of the electron displaces it from the C point of the furnace to the B point of the P/N junction, traveling CB space and producing **work** (T**)**. The force that moves the mass **m** along a space **(e)** produces work T. The consequence of the ICR collision against the P/N junction is its rupture releasing **heat** (**Q**). Thus, the internal mechanism of the heavy radioactive deuterotome changes work into heat, featuring the principle of 1st Law thermodynamics. Work T of 1J (Joule) produces or releases an amount of heat Q of 0,239 cal.

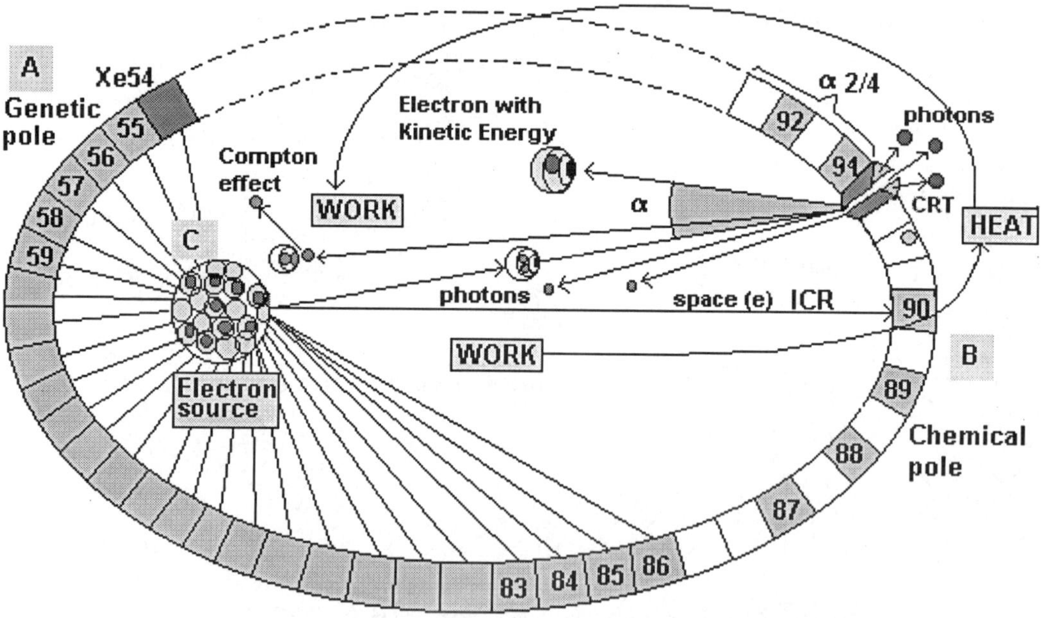

Fig. 52-b – **Heavy radioactive deuterotome -** Electron with kinetic energy moves along CB space (e) and produces work. This **work** is transformed into **heat** when the rupture of the TRC (Thermal Reserve Chamber) of the P/N junction occurs (First Law). The thermal photon (heat) released from the broken TRC strikes the intradeuterotomic electron. Part of the radiation energy goes to the electron. Thus, the **heat** released from the TRC produces **work** because it increases the kinetic energy of the electron (Second Law).

Thermal photons (heat) released by the rupture of the P/N junction are launched towards both the extradeuterotomic and the intradeuterotomic spaces. Electrons that lie in the electronic furnace also absorb the photons that were launched towards the deuterotome interior. Due to this photon absorption, the kinetic energy (Ke) of the electrons will rise and, consequently, its displacement and the production of work. Thus, the transformation of heat into work characterizes the fundamental of the 2nd Law of thermodynamics. A small calorie (1 cal) produces, in Joule's experiment, a work of 4.185 J. In vapor-driven engines, where there is a boiler and a condenser, the maximum theoretical output is 0.1 (10%). In explosion machines, where there is inflammable gas, the maximum theoretical output is 0.25 (25%). In the possible intradeuterotomic mechanisms of heavy radioactive elements, the rupture of the P/N junction, which comprises the TRC (Thermal Reserve Chamber), produces a true explosion, and is, therefore, the heat source that will produce work.

8.6.3 – Mass displacement effects in radioactive emissions

Emission of photon wave (gamma ray photon) and particles (alpha and beta) occur in radioactive degradations.

1) – Gamma ray photon emission (γ) - When a deuterotome emits a beta particle (β) the structure goes into an excited nuclear state and emits gamma ray photon, whose energy is 1.022 MeV/c². The emitting radioactive element recoils and goes into resonance; the photon emitted is captured by the absorbing deuterotome, which also recoils, moves from the initial position and also goes into resonance. This effect is known as *Mössbauer's effect.*

2) – Alpha particle emission (α2/4) - At the free extremity of the radioactive deuterotome chemical pole, is a proton-neutron-proton-neutron sequence (P/N/P/N), whose binding energy per nucleon (Ben) in U92/238 is 7.571 MeV. The shock of the ICR against the deuterotome wall provokes the rupture of the junction between the N-144 and the P-91 (see Figs. 52 and 52a) and releases alpha particles (α2/4) that move with a kinetic energy (Kα) of 5.696 MeV. This emission causes a recoil in the emitting deuterotome that is greater than the deuterotome producing gamma ray photon because the alpha particle has mass (α2/4 = 4.001506 u).

3) – Beta particle emission (β) - Beta decay takes place after one of the following conditions has occurred: 1) capture of a fast neutron; 2) emission of an alpha particle (alpha decay). The feature of the phylogenetic code of radioactive families (Ra88, Ac89, U92/235, and Th90) shows the successive betaemissions to balance the total energy of the system, preparing deuterotomes for alphaemission in the process of radioactive degradation. The same physical recoil phenomenon occurs in the deuterotome releasing beta particles through on a smaller scale, because the electron has smaller mass.

8.7 – The possible points of nuclear fission of uranium (U92/235)

Should the bindings of the deuterons that exist between xenon (Xe54) and radon (Rn86) be series-equivalent, the physical volume-reduction phenomenon would occur and radon would be the genetic pole of uranium; as a result, there would be no "electronic furnace" and the radioactivity resulting from the emission of alpha particles (α2/4) would not exist; the emission of beta particles (β) and gamma rays (ã) would then take place for the said radiations do not depend on the "furnace". If radon were the genetic pole of uranium, it would likewise be so for francium (Fr87), radium (Ra88), actinium (Ac89), thorium (Th90) and protactinium (Pa91). Because the bindings between deuterons from xenon to radon are parallel-equivalent type, the volume-reduction phenomenon does not take place. Instead, there is a concentration of the 32 electrons of this series in an agglomerate close to the genetic pole composed of xenon, called "electronic furnace", due to its electric load of 32 eV it is capable of ejecting electrons (electrons-projectiles) to strike the wall of the deuterotome chemical pole (target wall). Therefore, all the transuranium deuterotomes keep up the same structure that set the radioactivity mechanism in motion.

The fission of uranium-235 (U92/235) deuterotomes produced by atomic projectiles (slow neutrons) determines the formation of two elements of smaller deuterotomic numbers. Uranium-235 is not fissile at any point or any pole. The fission area is neither unspecific nor random. The phylogenetic lineage of deuterotomes shows that the atoms of hydrogen and deuterotomes have a particular set-up due to the synthesis mechanisms and the physical phenomena that actuate on the deuterons that make up deuterotomes, such as that of volume-reduction, the formation of the "electronic furnace" etc. As will be seen, uranium-235 has four vulnerable points for the neutrons that act as atomic projectiles. The points vulnerable to the action of neutrons-projectiles are located near the deuterotome's genetic pole, because the fissions happen in one of the proton-neutron (p/n) links designated by D54/D55, D55/D56, D56/D57 and D57/D58, as shown in the photo below.

Fig. 53. Simplified figure of uranium-235 deuterotome, in ring and elliptical shape with the genetic pole composed of xenon (Xe54) deuterotome. The clip indicates the point of rupture between Xe54 and deuterium D55 when uranium is bombarded by slow neutrons. The rupture at that point produces: U92/235 ———>Xe54 + Sr38 + 2 n.

The nuclear fission of uranium-235 can also occur between D55 (Cs55) and D56; between D56 (Ba56) and D57; between D57 (La57) and D58. The three black arrows in the illustration of uranium-235 below point out the possible points of nuclear fission when bombarded by slow neutrons.

Fig. 54. Three-dimensional figure of uranium-235 deuterotome in ring and elliptical-shapes, whose genetic pole is composed of xenon (Xe54) noble gas. Arrow 1 indicates one of the possible points of nuclear fission between deuterons D55 and D56 causing the nuclear reaction U92/235 ———> Cs55/141 + Rb37/93 + 2n; arrow 2 indicates the other possible fission point between D56 and D57 leading to the reaction U92/235 ——————> Ba56/140 + Kr36/93 + 2n; arrow 3 indicates the fourth possible fission point between D57 and D58 whose nuclear reaction is U92/235 ———> La57 + Br35 + 2n.

Cesium-141 deuterotome formed by fission goes through beta decay, thus forming barium-141 (Ba56/141) and carries on that mechanism until praseodymium-141 (Pr59/141) is formed. However, 0.03% of barium-141 acts differently; when it loses a neutron it forms barium-140 (Ba56/140); the latter, by way of the beta decay mechanism continues up to reach cerium-140 (Ce58/140). Rubidium-93 (Rb37/93) also goes through initial beta decay and forms strontium-93 (Sr38/93), which follows on with this mechanism until niobium-93 (Nb41/93) is formed. However, 1.4 percent of strontium-93 loses a neutron and forms one isotope of smaller mass number, that is, Sr38/92, which continues up to zirconium (Zr40/92) by beta decay. In every emission of beta particles (β), there is a concomitant emission of gamma rays. These reactions are symbolically shown below:

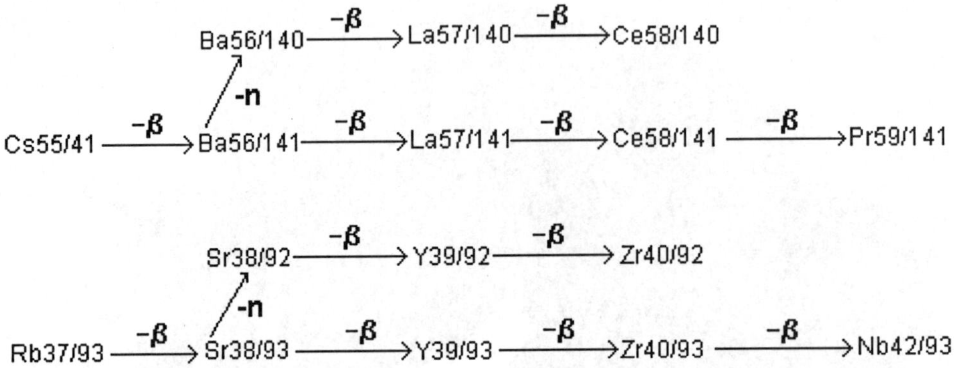

CHAPTER 9

Radioactive degradation of heavy deuterotomes

Radioactive degradation is a process that reduces both the number and mass of deuterotomes. Synthesis is a process that happens the other way round. How deuterotomic syntheses occur, however, are entirely different from how radioactive degradations do. In order for radioactive degradation to occur, a deuterotomic synthesis is initially necessary in keeping with a phylogenetic code. Several mechanisms participate in the synthesis process, which forms a number of elements from helium-4 (He2/4) to uranium-238 (U92/238). Synthesis can result from a simple process of capture or a dual process of capture and radioactive emission: a typical example of the former is the capture of alpha particles (α2/4) and the capture of neutrons plus the emission of beta particles (β) and gamma rays (γ) as examples of the latter.

The radioactive degradation of deuterotomes is a physical process; in other words, it does not depend on reactions for valences and is characterized by the emission of particles and photon waves (X rays and gamma rays). The emission of alpha particles reduces both the deuterotomic number and mass while the emission of beta particles only increases the deuterotomic number. The rise in the deuterotomic number can be regarded as a synthesis process because it changes a deuterotome of smaller number into one of greater number, increased by 1.

The radioactive property of uranium deuterotome (U92) and the elements that make up the radioactive families of radium, actinium, uranium-235 and thorium are not altered when subjected to high temperatures, pressure variations, electric loads, magnetic fields or other means. This fact is due to the steady constitution of deuterotome, which is composed of:

1) - A deuterons wall,

2) - A genetic pole composed of noble gas xenon (Xe54),

3) - The chemical pole formed by its six orbital electrons that corresponds to deuterons from D87 to D92, as is the case of uranium,

4) - An "electronic furnace" located inside the deuterotome, possibly one of the factors determining the radioactive property.

Nuclear fission alone is able to break apart the structure of uranium deuterotomes and other radioactive elements from thallium (Tl81) to protactinium (Pa91) and consequently, the "electronic furnace". The radioactive property of resulting deuterotomes is dependent on instability due to the excess of neutrons. Light fractions resulting from fission–rubidium-93 (Rb37/93), for instance–after emitting a beta particle (β) creates strontium-93 (Sr38/93) deuterotome, which emits a neutron and is changed into strontium-92; each one of these isotopes follows the series, emitting beta particles (β) until that the position of a stable deuterotome is reached.

9.1 – Atomic projectiles: Intra and extradeuterotomics

Two important atomic projectiles can be underlined, namely: the electron and the neutron. The electron is the intranuclear atomic projectile while the neutron is the extranuclear one. The electron, as electron-projectile, is the one that makes up the "electronic furnace" of intradeuterotome location; the electron of cathode rays functions as electron-projectile when ejected from the cathode (Crookes' tube) and hits the target wall on the opposite side. The neutron, as neutron-projectile, is produced in atomic reactors where it is accelerated and hurled against the target matter, such as uranium-238. It is, therefore, an atomic projectile of extradeuterotome location. In the figure below, a schematized model of deuterotome shows the area of performance of the said two types of atomic projectiles:

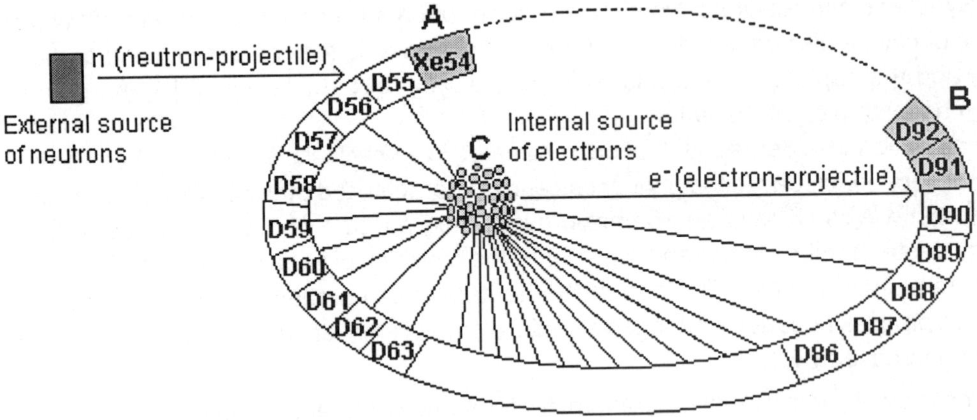

Fig. 55 - Illustration of heavy radioactive deuterotome (U92/235) showing the three major components: **A** – genetic pole composed of noble gas xenon (Xe54); **B** – chemical pole composed of the "alpha-appendix"; **C** – "electronic furnace" composed of the 32 electrons that make up the P-shell series occupying a position close to the genetic pole. See the electron-projectile configuration hurled from the "electronic furnace" against the target wall in the junction between deuterons D91 and D90 causing its rupture with consequent emission of alpha particles (α2/4); the neutron-projectile is thrown from the source of neutrons against the genetic pole in the D55/56 junction, causing nuclear fission.

The electron is a natural atomic projectile, either inside a heavy deuterotome that is a part of the "electronic furnace" or a part of the cluster of electrons forming cathode rays in Crookes' tube. It is naturally accelerated (with high speed ranging from 200,000 to 280,000 km/s). The 30,000 V electric discharge in the cathode of Crookes' tube ejects electrons against the opposite wall of the tube where it will interact with matter, thus producing fluorescence on the wall and emitting X rays; the four main interaction mechanisms are present in the intramural space of the tube wall:

1) - Photoelectric absorption,

2) - Compton effect,

3) - Formation of photon waves (X rays), and

4) - Formation of pairs of electron-positron.

Therefore, the cathode of Crookes' tube emitting electrons is a natural equivalent to a particles accelerator like the Betatron, the Synchrotron (28-GeV proton accelerator), the van der Graft particle accelerator or the Cyclotron, for example.

Likewise, a heavy radioactive deuterotome like uranium (U92) as well as deuterotomes with smaller number of protons up to thallium (Tl81), require a source of electrons with an electric load strong enough to eject and throw them against the target wall, similar to an accelerator of particles. Such being the case, the "electronic furnace" of a heavy deuterotome made up of 32 electrons lying at a point close to the genetic pole that has accumulated an electric load of 32 eV or 51.26 x 10^{-19} J, is a source of electrons in permanent activity that project particles against the wall, thus making up the deuterotome's chemical pole. The emission speed of alpha particles ($\alpha 2/4$) which happens in compliance with a certain pattern for a given deuterotome, determines its half-life. As far as half-life is concerned, uranium-238 (U92/238) and thorium-232 (Th90/232) are two sluggish radioactive deuterotomes, for uranium-238, when transformed into thorium-234, has a half-life of 8,10^9 years (eight billion years); thorium-234 half-life is far longer: 25, x 10^9 years (twenty-five billion years). Because both deuterotomes have very long deuterian structures, the shock of the electrons ejected by the "electronic furnace" against the target wall fails to have enough load to easily break apart the proton/neutron (p/n) binding between deuterons D91 and D90 of the chemical pole in those elements. In the series of uranium transformations (radium and actinium families) only uranium-234 (U92/234) has a long half-life of 3,10^6 years (three million years). However, the half-life of all the other elements is short; this fact suggests that the shorter the distance from the "electronic furnace" to the target wall, the easier the factor that facilitates the rupture of the proton/neutron bindings or, conversely, the greater the excited state in the electrons that make up the "electronic furnace", the greater the number of electrons that are thrown against the target wall situated in the chemical pole of the respective deuterotomes, which ever the case.

Radioisotopes with a very short half-life are quickly decayed towards inactive lead (Pb82); toward Pb82/206 for radium and actinium families; toward Pb82/207 for uranium-235 family and toward Pb82/208 for thorium family. Radium isotopes are, however, particularly important. Isotopes Ra88/222, Ra88/224 and Ra88/228 have all short half-lives of 16.4 days, 5.3 days and 9.67 years, respectively, while radium-226 (Ra88/226) has a half-life of 2,205 years. The radioactive property is not identical in every radioisotope: isotopes Ra88/222, Ra88/224 and Ra88/226 only emit alpha particles ($\alpha 2/4$) while Ra88/228 only emits beta particles (β). Radioisotopes Ra88/224 (alpha-emitter) and Ra88/228 (beta-emitter) are encountered in the radioactive family of thorium (Th90). Accordingly the phenomena resulting from the interaction of gamma rays with matter (which form the deuterotome wall) and the emission of photons are related to Ra88/228.

9.2 – Mechanisms of radioactive degradation: emissions of alpha (α2/4) and beta (β) particles.

The mechanisms leading to the radioactive degradation of heavy deuterotomes lie internally, and are simpler than the synthesis mechanisms of deuterotomes as a whole, which are external and dependent on the concentration of components.

The essence of radioactive degradation is encountered in alpha particles (α2/4) emission, causing the emitting deuterotome to lose two protons and two neutrons. Whenever a radioactive deuterotome emits a beta particle (β), the emitting deuterotome is increased by one proton, and the mass number is maintained. The neutron that has lost the electron acquires a proton-like status; because of that change the electromotive power is pushed down in every deuterium making up the series in the electrons orbital shell.

Radioactivity is a physical phenomenon directly associated with the following possible factors:

1) - Parallel equivalent bindings in the 32 deuterons forming the series in the electronic P-shell that follow after noble gas xenon (Xe54),

2) - The progressive formation of ring and elliptically-shaped structures from deuterium D55 to D86 with the electrons being consequently switched off that were in parallel equivalent bindings, all of which move towards a point close to the genetic pole to make up the "electronic furnace." Consequently, the physical volume-reduction phenomenon fails to happen and the structure formed by deuterons gets longer with round and elliptical shapes,

3) - Because the proton-electron span in the P-shell is very long–equal to 36 (6^2)–the 32 electrons in that shell are clustered near the genetic pole, thus making it possible for the electrons in the "electronic furnace" to lie relatively far from the chemical pole, acting as **distance of throw**,

4) -The "electronic furnace" starts to emit electrons of intranuclear location (electrons-projectiles); these electrons are accelerated and thrown against the deuterotome wall formed by the "alpha-appendix",

5) - The electrons thrown against the junctions in the pairs of deuteriums, formed by two protons and two neutrons, are enough to cause their rupture between deuterons D91/D90 and the emission of alpha particles (α2/4) with a great kinetic energy (Kα) ranging from 4 to 8 MeV. The typical equation in this reaction is as follows: U92/238 ———— >Th90/234 + α2/4,

6) - Electrons coming from the "furnace" and thrown against the wall of the chemical pole of the radioactive deuterotome bring about an excited nuclear state,

7) - The neutron that loses the electron and acquires the status of a proton causes a condition of intense excited nuclear state and emits gamma rays photon; the typical equation resulting from that transformation is as follows: Th90/234———> Pa91/234 + β^- + γ.

9.3 – Beta decay in nuclear fission

Beta decay itself is a simple process as well, albeit dependent on a previous structured based on one of the fundamental properties from the deuterotome genetic pole; in other words, its capability to adsorb neutrons and keep them stored on its surface and yield them to the post-noble gas deuterian structure as necessary to reestablish the balance in the system's electric loads. For instance, during the nuclear fission reaction where U92/235 upon being struck by a slow neutron, is split into two fraction-elements, which may be cesium-141 (Cs55/141) and rubidium-93 (Rb37/93) U92/235–the most typical. Cesium's genetic pole has eight neutrons adsorbed on its surface. The equation is as follows:

U92/235 + n —————————>Cs55/141 + Rb37/93 + 2n

where the genetic pole of cesium-141 can be represented by Xe54/132 (+ 8n).

Beta decay processes are encountered in heavy fractions, which held onto the original genetic pole, as well as in light fractions which set up their own genetic pole due to the electric loads being balanced once again, both resulting from the nuclear fission of uranium-235 (U92/235). Upon being struck by a slow neutron the latter goes through fission and creates two fraction-elements. The heavy fraction contains the original genetic pole of uranium's deuterotome and the light fraction–through a balance in the electric loads–creates its own genetic pole. Basically, the balance in electric loads results from the formation of series-equivalent bindings of deuterons numbered from D56 to D91 in uranium deuterotome. The atom of deuterium designated as D92 occupied a position of the post-noble gas in the deuterotome of rubidium-93. Thus, deuterons in the D56 and D91 sequence are series-equivalent connected and become electrically neutral (1 eV in each deuterium), going through the physical phenomenon of volume reduction. Therefore, by assuming both the electric and functional features of a genetic pole–formed by noble gas krypton (Kr36)–this unit then acquires the common property found in genetic poles, that is, the capability to adsorb neutrons on its surface.

As happens to cesium, rubidium-93 (Rb37/93) will have eight neutrons adsorbed in its genetic pole, in this case, noble gas krypton (Kr36), and can also be represented as follows: Kr36/84 (+ 8n) showing that eight neutrons are adsorbed by the genetic pole, the capacitance of which is significantly great.

The instability of cesium-141 (Cs55/141) and rubidium-93 (Rb37/93) results from the excess of the neutrons adsorbed by the surface of the genetic pole of each one of them, formed by xenon (Xe54) for cesium and by krypton (Kr36) for rubidium. Following the nuclear fission, depending on the proton-neutron (p/n) junction attained by the neutron, the fraction-elements will be different. The elements that may come up as products resulting from fission are mentioned below. Under each of those elements (heavy and light fractions) both the genetic pole that makes them up and the number of neutrons (in parentheses) are referred to that are adsorbed by the surface of the corresponding genetic pole:

$$\text{U92/235} + \text{n} \text{ -------------> } \frac{\text{Xe54/141}}{\text{Xe54/132 (+10n)}} + \frac{\text{Sr38/93}}{\text{Kr36/84 (+6n)}} + 2\text{n}$$

$$\text{U92/235} + \text{n} \text{ -------------> } \frac{\text{Cs55/141}}{\text{Xe54/132 (+8n)}} + \frac{\text{Rb37/93}}{\text{Kr36/84 (+8n)}} + 2\text{n}$$

$$\text{U92/235} + \text{n} \text{ -------------> } \frac{\text{Ba56/141}}{\text{Xe54/132 (+6n)}} + \frac{\text{Kr36/93}}{\text{Kr36/84 (+10n)}} + 2\text{n}$$

$$\text{U92/235} + \text{n} \text{ -------------> } \frac{\text{La57/141}}{\text{Xe54/132 (+4n)}} + \frac{\text{Br35/93}}{\text{Kr36/84 (+12n)}} + 2\text{n}$$

Because noble gases are turned into armored and compact structures (through the physical phenomenon of volume-reduction) all neutrons then adsorbed fail to penetrate their structures and are therefore, stored externally and likely to be distributed in pairs in order to stabilize the deuterotome structure. The displacement in the **pair of neutrons** departing from the genetic pole and moving toward the post-noble gas deuterian structure causes thickness **d** of the dielectric to be widened and both the unitary and global capacitance to be decreased. From one of the neutrons that make up the pair, an electron (beta particle) is emitted and the neutron acquires a proton-like status, and the deuterotomic number goes up. The sequence of beta decay in cesium-141 can be shown as follows:

$$\frac{\text{Cs55/141}}{\text{25s}} \xrightarrow{-\beta,\ \gamma} \frac{\text{Ba56/141}}{\text{18m}} \xrightarrow{-\beta,\ \gamma} \frac{\text{La57/141}}{\text{4h}} \xrightarrow{-\beta,\ \gamma} \frac{\text{Ce58/141}}{\text{325d}} \xrightarrow{-\beta,\ \gamma} \frac{\text{Pr59/141}}{\text{Stable}}$$

The half-life (t ½) of each element is expressed in seconds (s), minutes (m), hours (h) and days (d), under each one of the elements thus formed. Emissions of gamma rays are accompanied by emissions of beta particles (β).

The transfer of the pair of neutrons from the genetic pole to the deuterian structure may occur as illustrated in the following scheme.

Thus, the genetic pole of cesium progressively distributes pairs of neutrons toward the deuterian structure, and an electron (beta particle) is emitted by one of them, being transformed into a proton. This process goes on until the supply of neutrons adsorbed in the genetic pole is depleted.

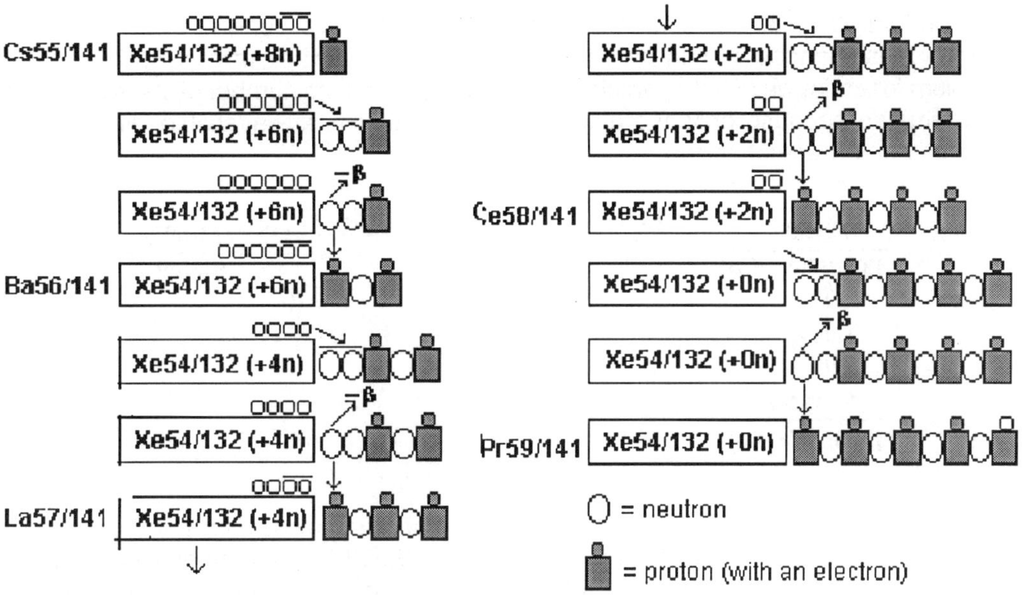

As for rubidium-93 (Rb37/93), the physical phenomenon of neutrons adsorption through the surface of the genetic pole–the case of noble gas krypton (Kr36)–occurs in the same pattern as cesium. Krypton is also capable of storing eight neutrons on its surface. The sequence of transformations is similar to that of cesium: only the deuterotomes then formed will be different.

$$\frac{\text{Rb37/93}}{\text{6s}} \xrightarrow{-\beta,\ \gamma} \frac{\text{Sr38/93}}{\text{7m}} \xrightarrow{-\beta,\ \gamma} \frac{\text{Y39/93}}{\text{10h}} \xrightarrow{-\beta,\ \gamma} \frac{\text{Zr40/93}}{1.5.10^{6}\ \text{y}} \xrightarrow{-\beta,\gamma} \frac{\text{Nb41/93}}{\text{Stable}}$$

The mechanism that governs the emission of beta particles (β), after the pair of neutrons has been transferred from the genetic pole to the deuterons structure, is similar to that of cesium-141. The same thing happens to all the other fractions, light and heavy.

The stability of niobium (Nb41/93) and praseodymium-141 (Pr59/141) deuterotomes results from the number of neutrons being depleted which were adsorbed in their respective genetic poles. However, 0.03 percent of Ba56/141 loses one neutron and originates an isotope of smaller mass number, namely, Ba56/140. Accordingly, there will only be five neutrons adsorbed on the surface of the genetic pole. This reduces the number of emissions of beta particles (β) because only two pairs of neutrons will be available to be used. The sequence of transformations is as follows:

$$\frac{\text{Ba56/140}}{\text{[Xe54/132(+5n)]}} \xrightarrow{-\beta,\ \gamma} \frac{\text{La57/140}}{\text{40h}} \xrightarrow{-\beta,\ \gamma} \frac{\text{Ce58/140}}{\text{Stable}}$$

$$\uparrow -\text{n}$$

$$\frac{\text{Ba56/141}}{\text{[Xe54/132(+6n)]}}$$

Strontium-93 (Sr38/93) also loses one neutron. However, only 1.4 percent of that deuterotome is changed into an isotope of smaller mass number, that is, Ba56/140. The mechanism for emission of beta particles (β) is the same and the number of transformations is also reduced because only two pairs of usable neutrons are available. The scheme below illustrates these changes.

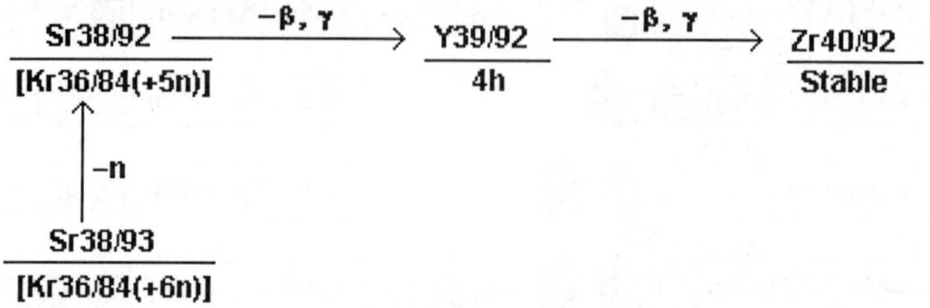

Uranium-238 (U92/238) is not fissile. When bombarded by neutrons, it captures that particle and increases its mass number, originating U92/239. When a neutron enters the "alpha-appendix" the capacitance is unbalanced and the deuterotome consequently emits a beta particle (β), which forms neptunium-239 (Np93/239); the latter also emits a beta particle (β) being changed into plutonium-239 (Pu94/239). The emission of an alpha particle (α2/4) by plutonium, for its turn, produces uranium-235 (U92/235). The following illustration shows all of these modifications:

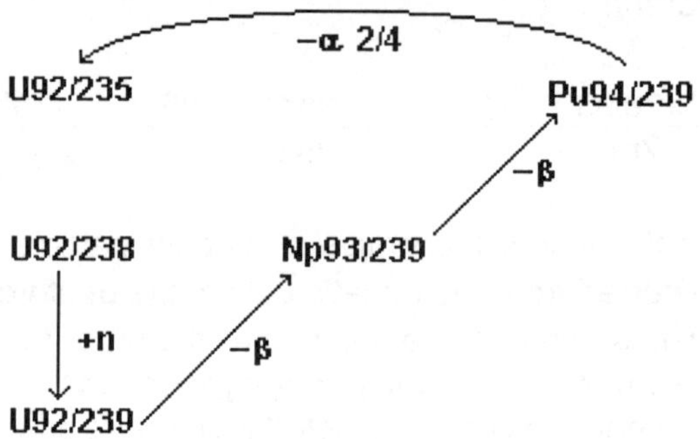

9.4 – Alpha and beta decays in natural radioactive degradation

A heavy deuterotome like that of uranium, for example, can be schematically represented as shown below, where **A** is the genetic pole, **B** the chemical pole and **C** the "electronic furnace." The process of alpha and the beta decay occurs in the chemical pole. The scheme from (a) to (e) is a microscopic sample of the transformations that happen in the chemical pole where the "alpha-appendix" is located. In the sequence, there is one emission of alpha particle (α 2/4), two emissions of beta particles (β) and one other emission of an alpha particle.

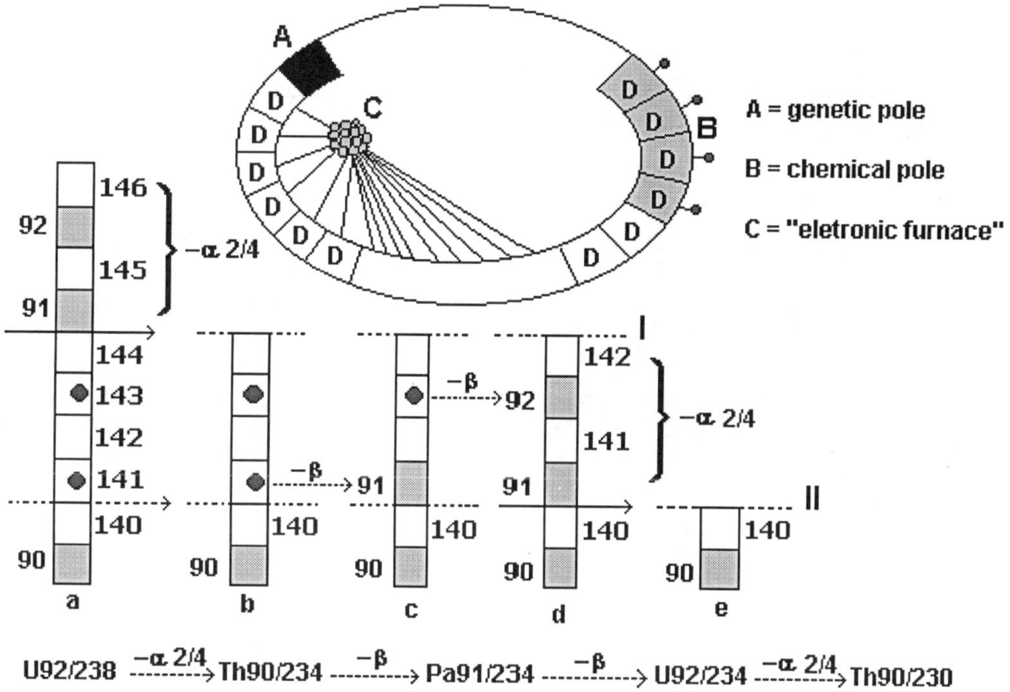

Fig. 56. Chemical pole of U92: changes that occur upon the emission of alpha and beta particles from U92/238 deuterotome during the radioactive degradation of radium and actinium families (partial scheme). 1) - Emission of alpha particles formed by deuterons D91/D92 due to a rupture in the proton/neutron binding at **point I.** 2) - Beta particles emission by neutrons 141 and 143 showing protons 91 and 92 in b/**c**. In **d**, the alpha-unit is ready to be emitted to originate Th90/230

Fig. 57 - Spatial illustration, in ring and elliptical shapes, of uranium deuterotome (U92) showing the genetic pole constituted by xenon (Xe54). Deuterons D55, D56 and D57 show the deuterons sequence that make up the deuterotome. Close to the genetic pole is an agglomerate of 32 electrons that form the "electronic furnace", electrons that range from D55 (Cs55) to D86 (Rn86).

Fig. 58 - Spatial illustration of uranium deuterotome (U92) viewed from the chemical pole (constituted by deuterons) that range from D92 to D87, representing the four or six possible free valences of uranium. In the center and in the interior of the structure are shown the electrons that compose the "electronic furnace".

The location of the "electronic furnace" underlying the genetic pole provides space for electrons-projectiles to be thrown against the proton/neutron (p/n) junctions of the pair of deuterons at the free end of the chemical pole, causing its rupture and consequent emission of alpha particles ($\alpha 2/4$). The next scheme aims at showing that the distance between the genetic pole and the "electronic furnace" is smaller than the latter and the chemical pole. Thus, the scheme of a heavy deuterotome resembles that of Crookes' tube, where the "furnace" corresponds to the cathode, the chemical pole to the target wall of the tube and the electron-projectile of the "furnace" to the cathode rays.

9.5 – Radioactive degradation in the radium, actinium, uranium-235, and thorium families.

In the radioactive degradation in radium and actinium families, uranium-238 (U92/238) is the initial deuterotome, there being emissions of alpha particles ($\alpha 2/4$) from both of them, which are changed into Th90/234. In both families the emission of beta particles (β) from thorium-234, transforms it into protactinium-234 (Pa91/234). An emission of beta particles by this deuterotome transforms it into U92/234. This deuterotome, upon emitting an alpha particle ($\alpha 2/4$) originates thorium-230 (Th90/230).

From Th90/230, the degradation patterns turn out to be different for the two families (radium and actinium). In the radium family, the deuterotome starts emitting four alpha particles ($\alpha 2/4$) successively, being changed into lead-214 (Pb82/214). In the actinium family, Th90/230 still emits beta and alpha particles alike and is changed into an isotope of smaller mass number, namely, Th90/226. From the latter, four successive emissions of alpha particles occur, originating lead-210 (Pb82/210). Both Pb82/214 and Pb82/210 make two successive emissions of beta particles, being changed into bismuth-214 (Bi83/214) and polonium-214 (Po84/214) in the former, and bismuth-210 (Bi83/210) and polonium-210 (Po84/210) in the latter. The overall transformations can be seen in the table of radioactive degradation in the radium and actinium families.

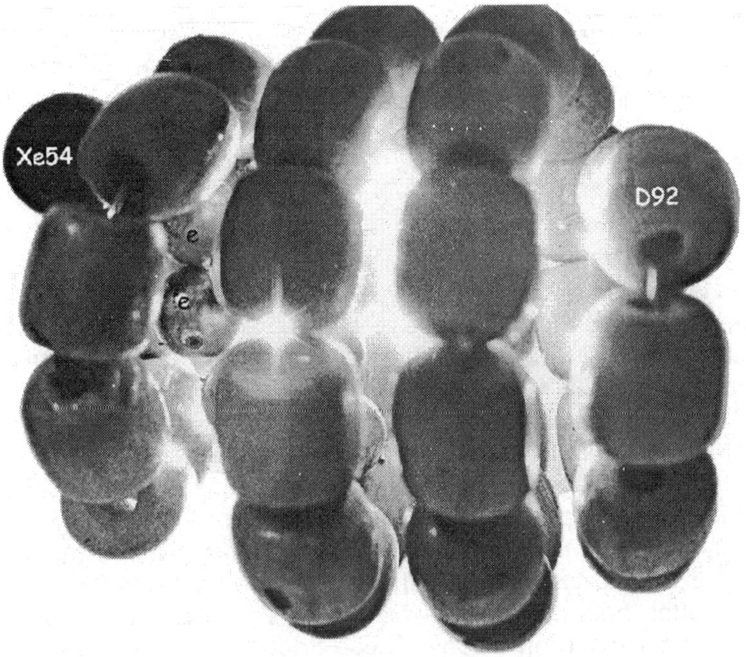

Fig 59 – Spatial representation in round and elliptical shapes of the structure of uranium deuterotome (U92) somewhat stretched out, with its chemical pole on the right (D92) where chemical reactions and radioactive degradations occur. The genetic pole lies on the left, represented by noble gas xenon (Xe54) where the junctions are located subject to ruptures due to the action of slow neutrons acting as atomic projectiles. The "electronic furnace" has an intradeuterotomic location, very close to the genetic pole. A few electrons are shown inside the deuterotome, making up the "furnace."

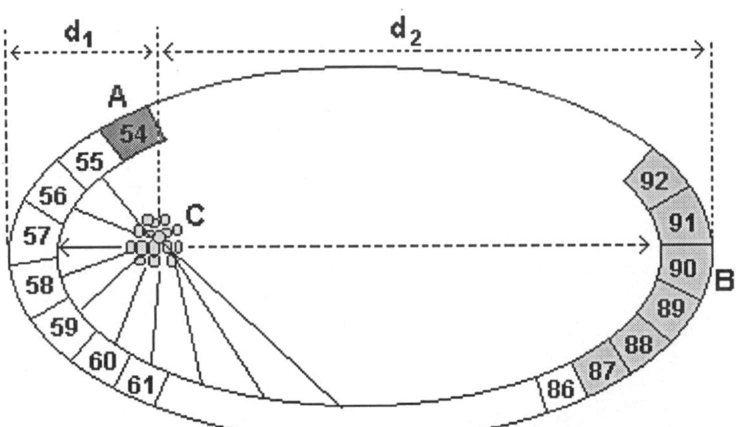

Fig. 60 - Illustration of uranium deuterotome (U92) structure, showing the genetic pole (**A**), the chemical pole (**B**) and the "electronic furnace" (**C**), the latter lying close to the genetic pole. See that distance d1 between the genetic pole and the "furnace" is far shorter than distance d2, between the latter and the chemical pole. The electron-projectile uses stretch CB to reach D91/D90 junction, breaking it, and emitting alpha particle (α2/4).

The radioactive degradation in the thorium family is particularly striking: it begins from thorium-232 (Th90/232) and not from uranium (U92). This results from two basic conditions: 1) during the radioactive degradation of U92/238 and U92/235, the mediums do not go past Th90/232; 2) during the process of deuterotomes synthesis (see table of possible synthesis mediums), the latter goes through a stop upon reaching the composition of Th90/232. Thus, Th90/232 maintains the same deuterotomic structure as uranium with a genetic pole, a chemical pole and an "electronic furnace", provided with the same radioactive properties. Thorium-232 emits alpha particles and changes into radium-228 (Ra88/228). The latter, due to two successive emissions of beta particles is changed into Th90/228. From that isotope four successive emissions of alpha particles (α2/4) occur, and the condition of Pb82/212 is reached. This element, by means of three successive emissions of beta particles, is transformed into Bi83/212, Po84/212 and At85/212, all of them isobars. The final stage in the radioactive degradation of Th90/232 is inactive lead-208 isotope (Pb82/208). Pb82/207 and Pb82/206, both inactive, are the final elements deriving from the degradation of U92/235 and U92/238, respectively. Below are the schemes of radioactive degradation in the four families, throughout their stages. Where an element emits alpha and beta particles there are two fractions–alpha and beta (e.g.: bismuth-alpha fraction and bismuth-beta fraction)–which indicates that the fraction first emits alpha and, after, beta particles or vice-versa, since both move toward the same deuterotome right after these emissions. These fractions can be seen in bismuth and actinium.

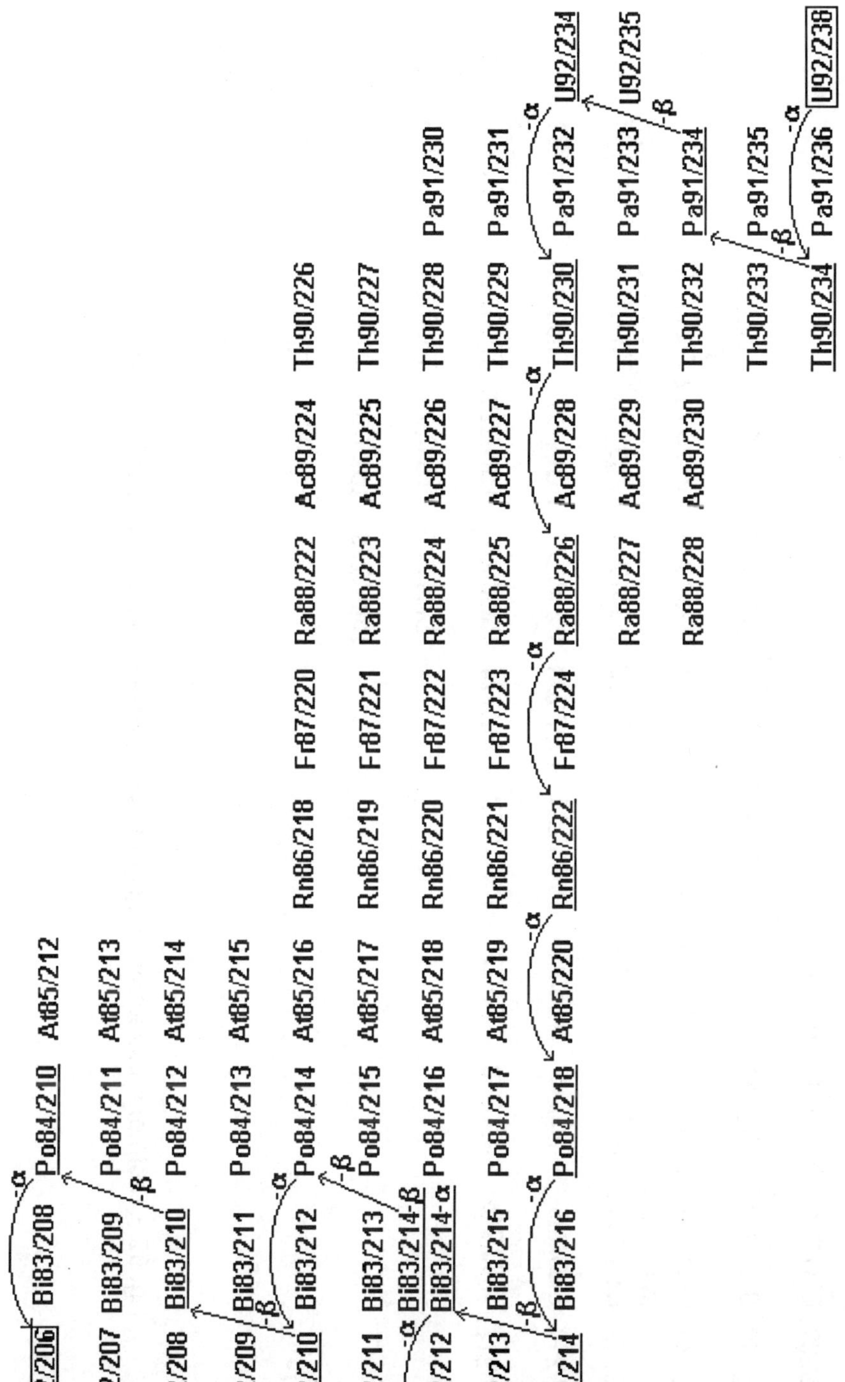

Fig. 61 - Table of radioactivite degradation sequence of radium family (Ra88)

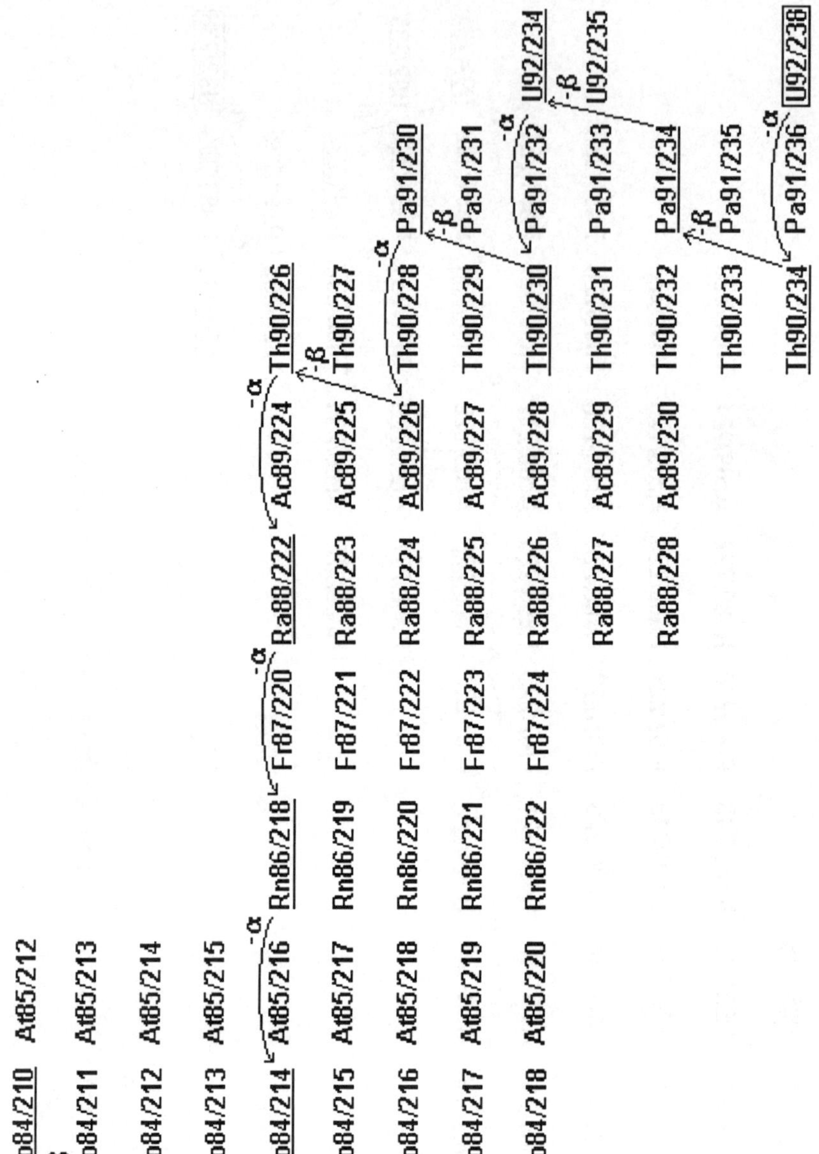

Fig. 62 - Table of radioactive degradation sequence of actinium family (Ac89)

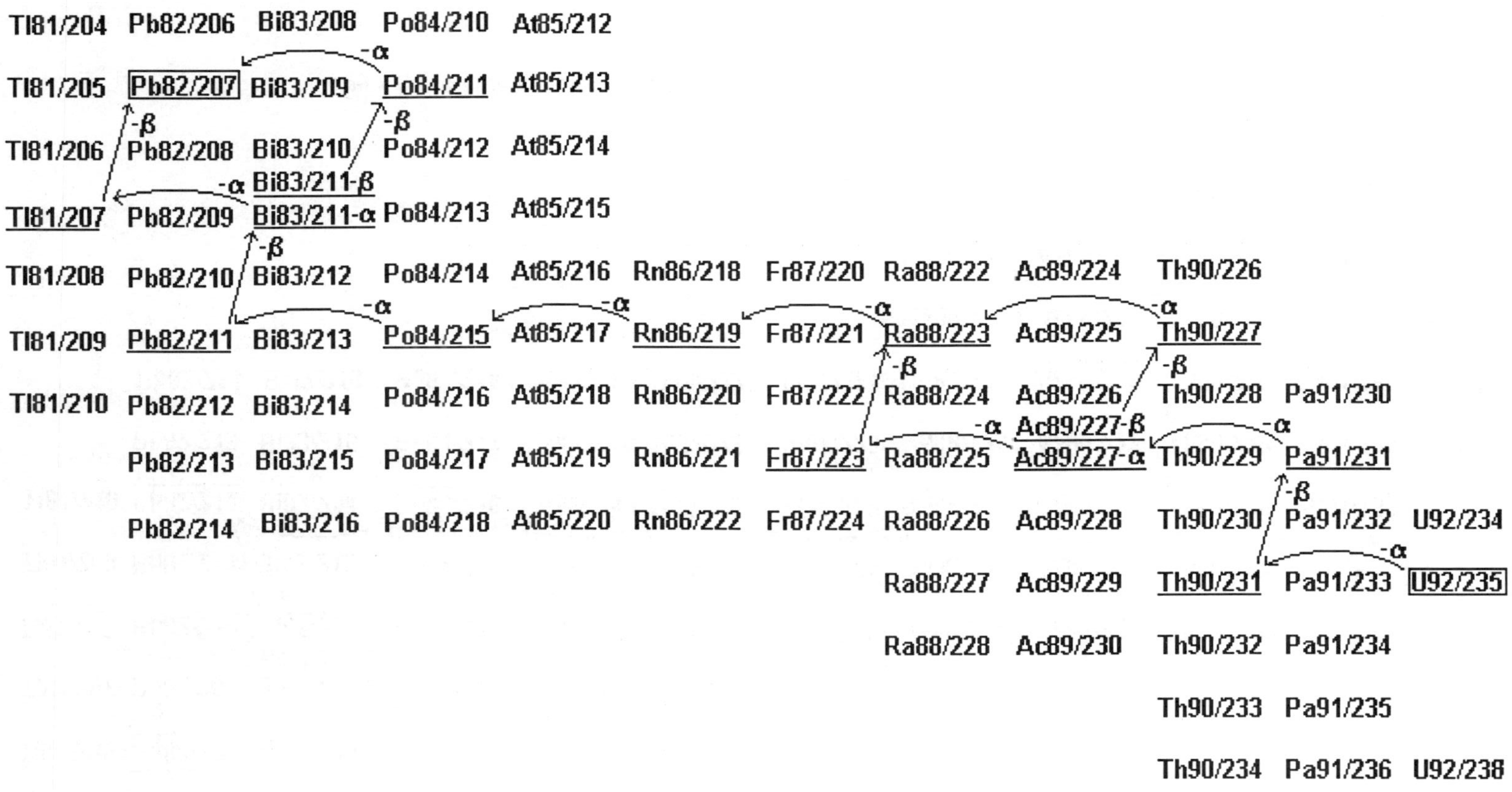

Fig. 63 - Table of radioactive degradation sequence of uranium-235 family

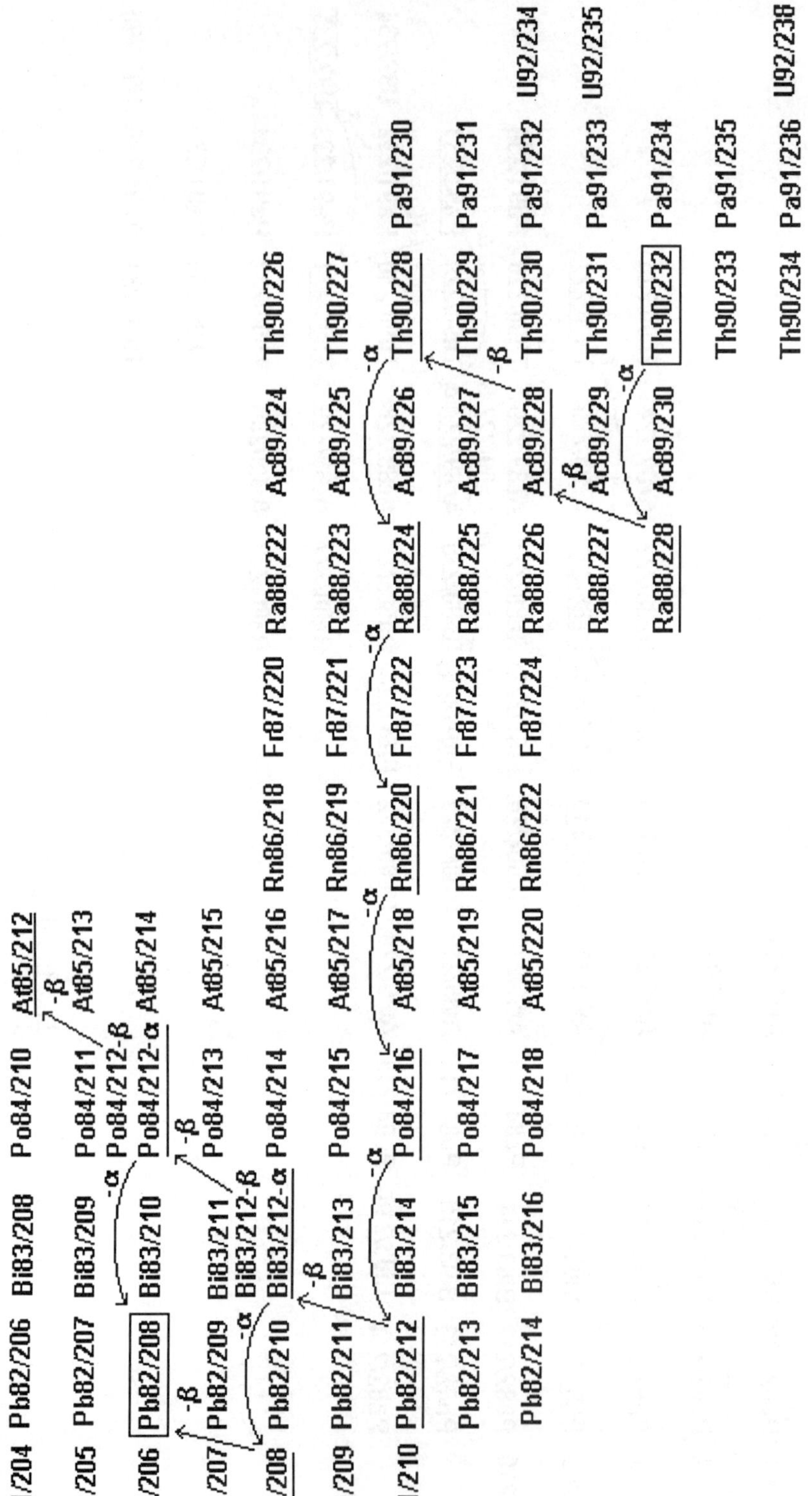

Fig. 64 - Table of radioactive degradation of thorium family (Th90)

9.6 – Phylogenetic code of "alpha-appendix" in radioactive families

The mechanisms that capture neutrons (n) and alpha particles (α2/4) and the one that emits beta particles (β) are present during the process of deuterotomes synthesis (nucleosynthesis). Conversely, the mechanisms that trigger emissions of alpha (α2/4) and beta (β) particles are present during the process of radioactive degradation in heavy elements.

In deuterotomes with a capacitive series-equivalent structure (up to Xe54), where there is no "electronic furnace" and the deuterotomic structures are linear and short and the spatial arrangement fails to take on a typically spiral-shaped form, there is no radioactive degradation; in other words, no emission of alpha particles (α2/4).

In heavy deuterotomes with capacitive parallel equivalent bindings, post-xenon (Xe54), the "electronic furnace" acquires a defined physical set-up, the deuterotomic structures are both linear and long and the spatial arrangement assumes a typically wound-like spiral-shaped form. Radioactive degradations from uranium (U92) to thallium (Tl81) occur in these elements due to alpha and beta decays.

9.6.1 – Basic factors that determine radioactivity

The phylogenetic code stored in the "alpha-appendix" of heavy deuterotomes depends most likely on a synthesis mechanism set in motion from thallium (Tl81) with mass numbers equal to 203, 204 and 205. The three major thallium families are as follows:

1) - Thallium-203 (Tl81/203) that originated radium and actinium families, both radioactive,

2) - Thallium-204 (Tl81/204) that originated the radioactive family of uranium-235,

3) - Thallium-205 (Tl81/205) that originated the radioactive family of thorium.

Natural radioactivity is in no way a phenomenon that occurs at random, in other words, it does not happen by accident without a defined stimulus. Because it is a physical phenomenon, it is subject to certain basic factors some of which are self-evident:

1) - An alpha or beta particle cannot be emitted until the structure of the deuterotome's chemical pole is physically and electrically ready,

2) - During the process of natural radioactive degradation from uranium (U92) to lead (Pb82), there is no emission of neutrons, but rather alpha and beta particles as well as gamma radiations,

3) - The capacitance of ensuing deuterotomes is slower after an alpha decay and higher after beta decay,

4) - The binding energy per nucleon (ben) in heavy and radioactive deuterotomes is slower than that one found in non-radioactive elements,

5) - In the "alpha-appendix" of heavy deuterotomes, the protons and neutrons should occupy a defined position–different for each radioactive family,

6) - During the processes of natural radioactive degradation in the four families (radium, actinium, uranium-235 and thorium), bismuth (Bi83) acts as though an amount of that element consisted of one alpha and one beta fractions, that is, bismuth-alpha and bismuth-

beta respectively, an indication that a fraction of that element (alpha fraction) first emits an alpha particle (α2/4) and then a beta particle (β). The other fraction (beta-fraction) does just the opposite: first it emits a beta particle and then an alpha particle,

7) - In uranium-235 family, two actinium fractions are encountered, namely alpha-actinium and beta-actinium,

8) - In the linear structure of radioactive elements of the four families, protons and neutrons are arranged as though they were "alpha particles" with the following even numbers of neutrons in their compositions: α2/4, α2/6, α2/8 and α2/12, thus determining a capacitive value that gets smaller and smaller for larger number of neutrons,

9) - The hypercapture of neutrons, inserted by beta decay, from thallium-203, 204 and 205 families,

10) - Availability of a phylogenetic code in the composition of the "alpha-appendix" structure linked to mercury (Hg80/202, 203 and 204) deuterotomes.

9.6.2 – Phylogenetic codes in the "alpha-appendix"

The phylogenetic lineage in heavy radioactive and trans-mercury deuterotomes belonging to radium, actinium, uranium-235 and thorium families maintains an alpha-unit composition code along the "alpha-appendix."

The alpha-unit composition code in the "alpha-appendix" from Hg80/202 onwards, participating in the composition of radium radioactive family is as follows:

Fig. 65 – **Radium family**. Illustration of the "alpha-appendix starting from Hg80/202 showing the phylogenetic code in the alpha-unit composition. From the deuterotome mass that makes up Hg80/202, the following alpha-units are encountered in sequence: α2/12, α2/4, α2/4, α2/4, α2/8 and α2/4. The six neutrons that distribute electrons during the beta decay process are as follows: n-143, n-141, n-131, n-128, n-126 and n-124 (see Figs. 17, 61 and 69 that are complementary).

The alpha-unit composition code in the "alpha-appendix" from Hg80/202 onwards, participating in the composition of actinium radioactive family is as follows:

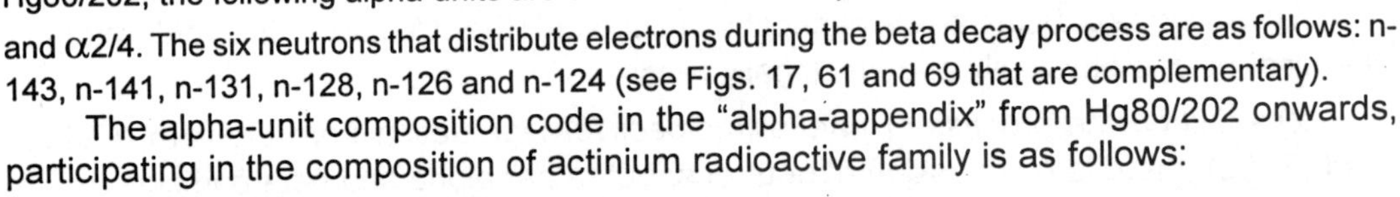

Fig. 66 – **Actinium family**. Illustration of the "alpha-appendix" starting from Hg80/202 showing the phylogenetic code in the alpha-unit composition. From the mass of deuterotome that makes up Hg80/202, the following alpha-units are encountered in sequence: α2/8, α2/4, α2/4, α2/4, α2/12, and α2/4. The full squares represent the protons; the blank ones, the neutrons and the blank ones with a central circle, the neutrons that emit one beta particle (β) and then be changed into protons. The neutrons that distribute electrons during the beta decay process are as follows: n-143, n-141, n-139, n-136, n-127 and n-124 (see Figs. 18, 62 and 70 that are complementary).

The alpha-unit composition code in the "alpha-appendix" from Hg80/203 onwards, participating in the composition of uranium-235 radioactive family is as follows:

Fig. 67 – **Uranium-235 family**. Illustration of the "alpha-appendix" starting from Hg80/203, showing the possible phylogenetic code in the alpha-unit composition in the following order starting from the mass of mercury: α2/8, α2/4, α2/4, α2/6, α2/6 and α2/4. The four neutrons that distribute electrons for the beta decay process are as follows: n-140, n-135, n-128 and n-125 (see Figs. 19, 63 and 71 that are complementary).

The alpha-unit composition code in the "alpha-appendix" from Hg80/204 onwards, participating in the composition of thorium radioactive family is as follows:

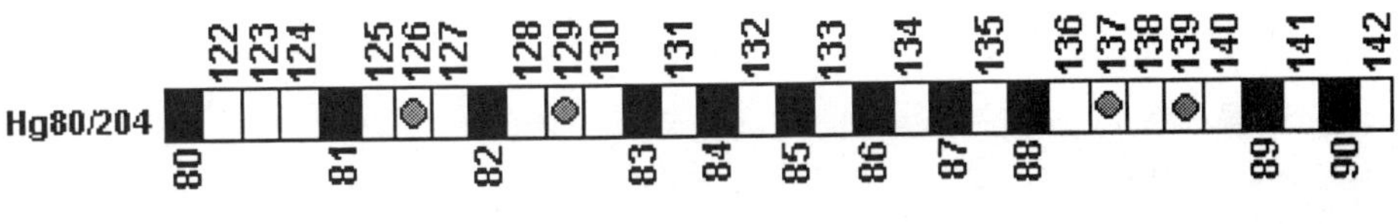

Fig. 68 – **Thorium family**. Illustration of the "alpha-appendix" leaving from Hg80/204, showing the phylogenetic code in the alpha-unit composition in the following order starting from the mass of mercury to the end: α2/8, α2/4, α2/4, α2/8 and α2/4. The neutrons that distribute electrons for the beta decay process are as follows: n-139, n-137, n-129 and n-126 (see Figs. 20, 64 and 72 that are complementary).

The sequence in the alpha-unit of heavy deuterotome's "alpha-appendix" previously referred to is summarized in the table below:

Radium family	Hg80/202	α2/12	α2/4	α2/4	α2/4	α2/8	α2/4
Actinium family	Hg80/202	α2/8	α2/4	α2/4	α2/4	α2/12	α2/4
U-235 family	Hg80/203	α2/8	α2/4	α2/4	α2/6	α2/6	α2/4
Thorium family	Hg80/204	α2/8	α2/4	α2/4	α2/8	α2/4	

Calculating the capacitance value for each alpha-unit, by means of **C equation**, produces the following results expressed in electron-farad (eF): α2/4, (1.5070), α2/6 (0.7854), α2/8 (0.3927) and α2/12 (0.2618).

9.6.3 – Diagrams showing the sequence of radioactive degradation in radium, actinium, uranium-235 and thorium families

The diagrams below show the process of alpha and beta decays in the radioactive degradation of radium and actinium (whose origin is U92/238), besides uranium-235 and thorium. The sequence in radioactive emissions is dependent on the phylogenetic code determined by the arrangement of protons and neutrons in the "alpha-appendix", post-mercury. The black squares represent the protons and related numbers, the "loci" initials of protons; the white squares represent the neutrons that participate in the composition of the alpha-unit–of variable number–depending on the radioactive family; the white squares with a circle in the center represent the neutron that distributes the electron for beta decay. The neutrons are numbered in increasing order from mercury up to the end, starting from n-122. Bismuth (Bi83) and actinium (Ac89) are represented twice in the diagrams due to alpha and beta fractions available in a large number of them.

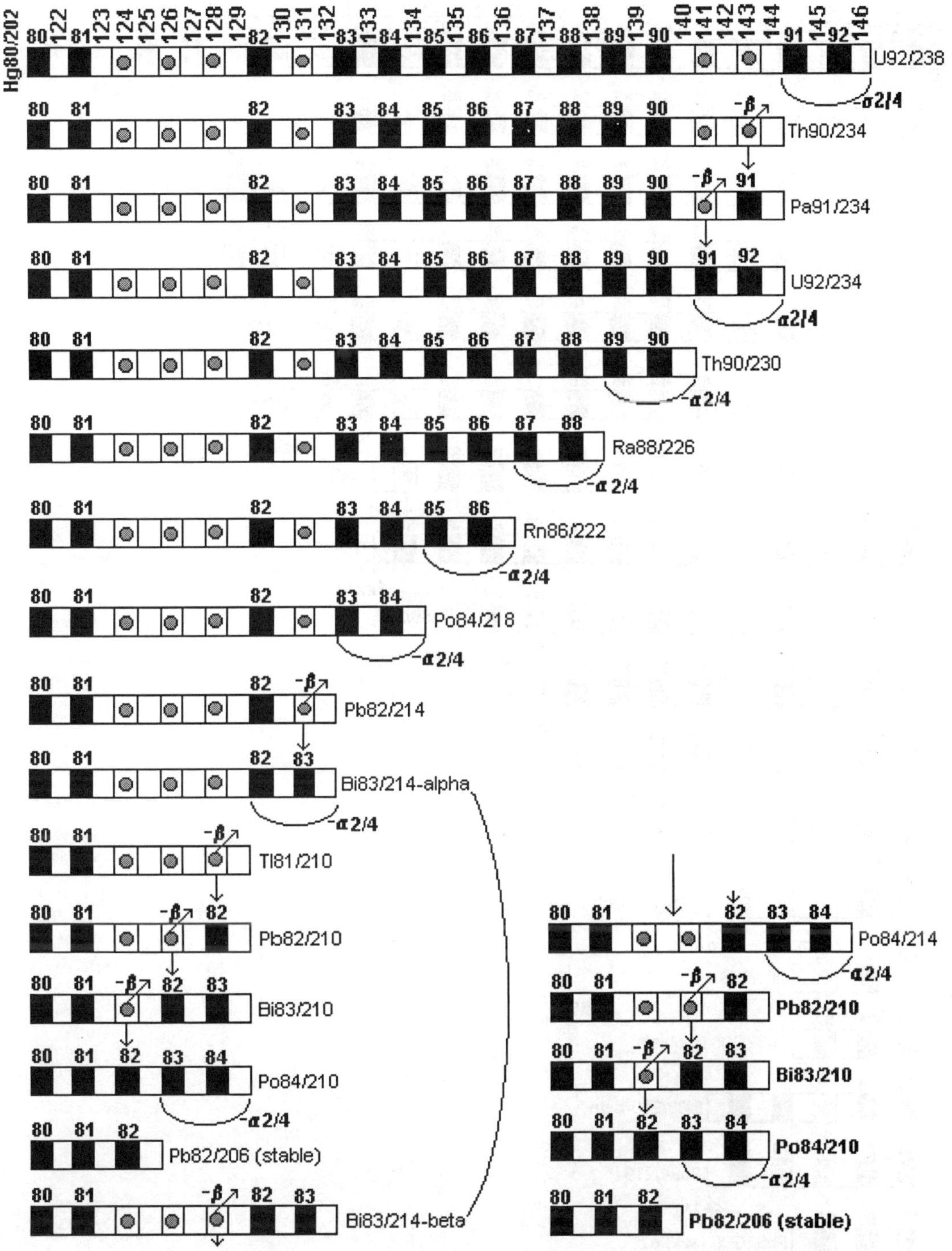

Fig. 69 - Radioactive degradation of radium family. The phylogenetic code in the "alpha-appendix" sets forth both alpha and beta decays. See the bismuth-alpha and bismuth-beta fractions, which prioritize emissions.

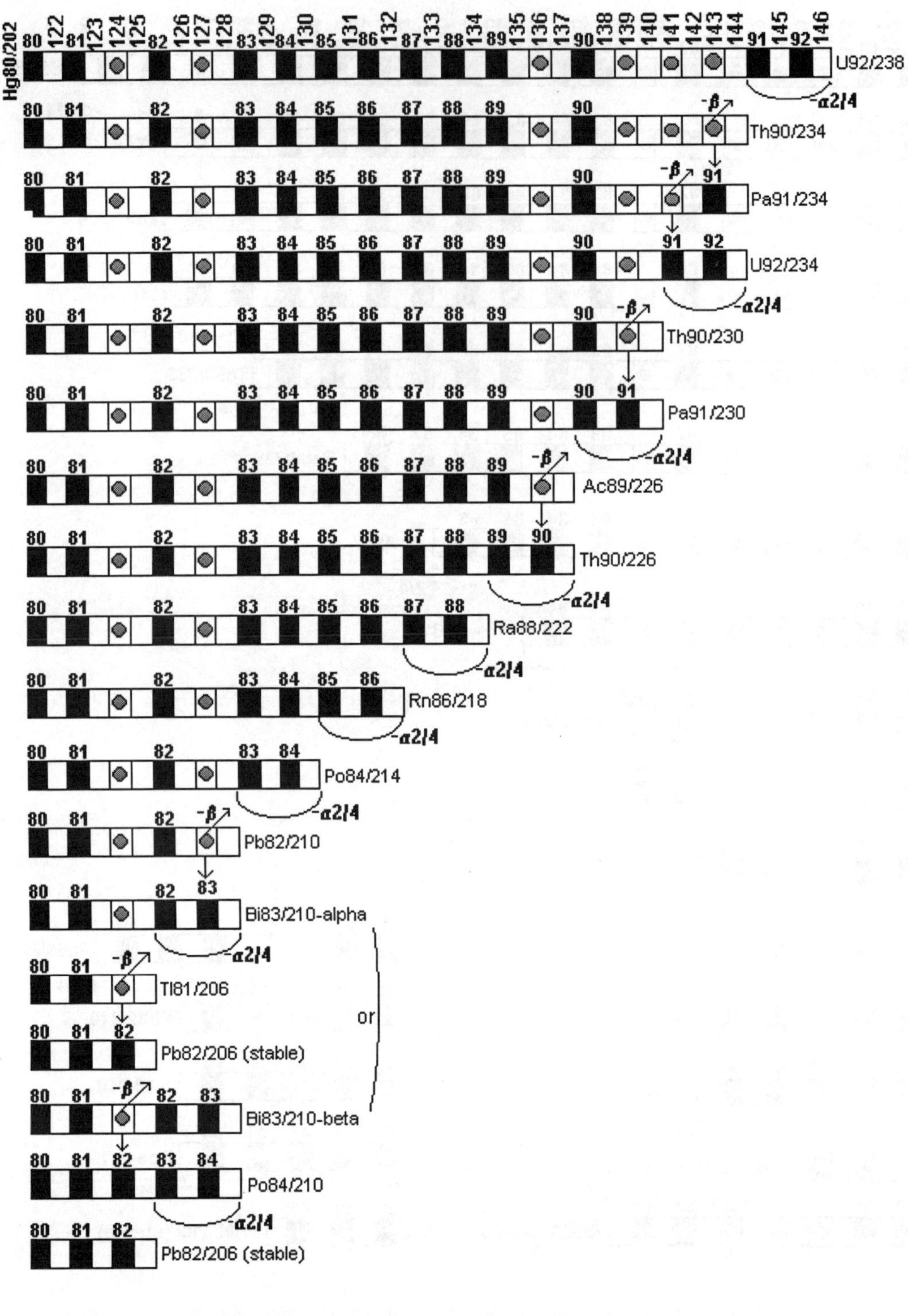

Fig. 70 – Radioactive degradation of actinium family. Mechanisms of degradation: from uranium-238. See bismuth-alpha fraction and bismuth-beta fraction, which prioritize one or other radioactive emission. Both, however, move toward stable Pb82/206.

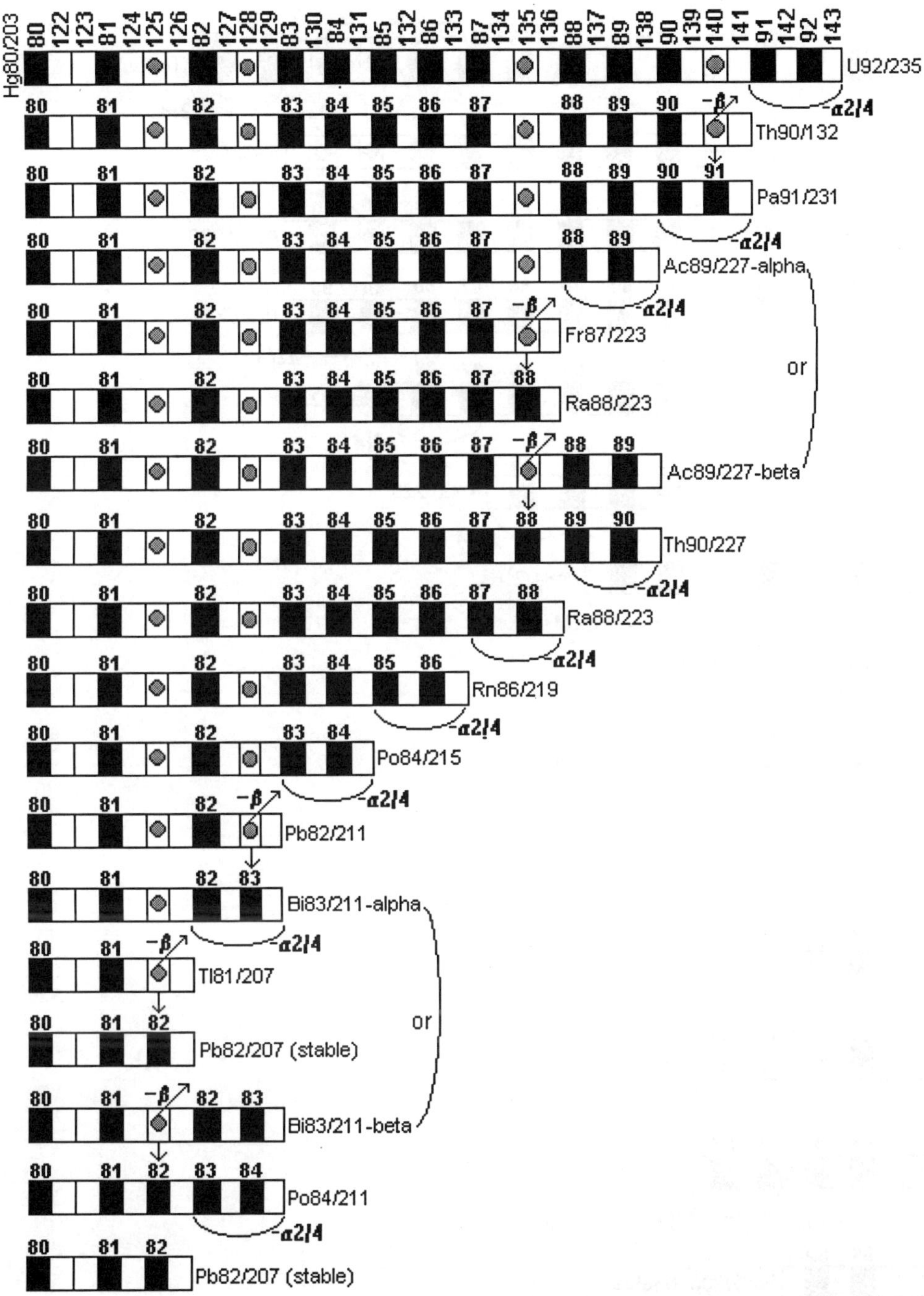

Fig. 71 – Radioactive degradation of uranium-235 family. Illustration of alpha and beta decays. See alpha and beta fractions of bismuth (Bi83) and actinium (Ac89), which prioritize the emission of one or other radiation. Pb82/207 is the final and stable element.

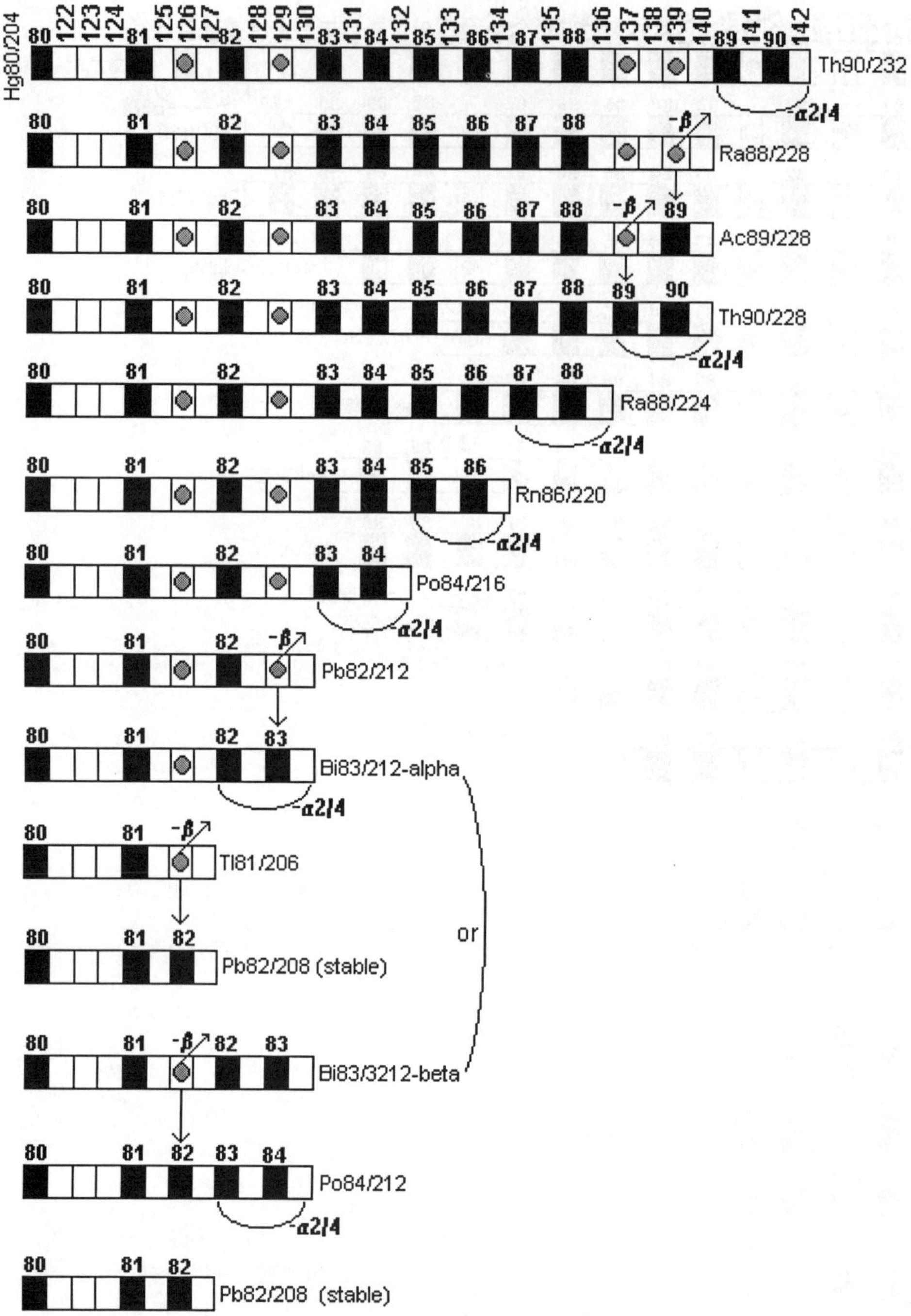

Fig. 72 – Radioactive degradation of thorium family: mechanism of degradation by alpha and beta decays. See the deuterotome of bismuth: there are alpha and beta fractions, each prioritizing one of the radioactive emissions. Pb82/208 deuterotome is stable in this series.

CHAPTER 10

Electrical and thermal conductivities

10.1 – Band theory of solids

Whether in the classic atomic theory or the one proposed in this study, the band (or strip) theory of electrical conductivity is not altered. An electric current flows on the surface of a conductor wire stemming from a source that distributes electrons, heading towards the receiving target (a resistor, for example). So that the flow of electrons can circulate freely on the surface of the conductor wire, a carry-over mechanism is required. In order to elucidate this carry-over mechanism, both the classic atomic theory and the quantum theory are called into play, the latter associated with electrons.

The band theory of solids (or strip theory) is backed by the two theories referred to above and turns out to be quite rational. It explains, for instance, why some crystals are better conductors of electric energy than others, why some can have an insulating property, why others are semiconductors and others resistors.

An overall account of the theory at issue, where reasoning can emerge, starts off by the atomic structure proposed by Rutherford-Bohr (1911-1913) and Thomson. The electronic crown that roams around the atom nucleus would be formed by electron shells or **atomic shells** and designated by K, L, M, N, 0, P and Q, each far from the other, the value of which is the square in the series of whole numbers from 1 to 7, respectively, 1, 4, 9, 16, 25, 36 and 49. A regular number of electrons can be found in each shell. Because the classic atomic theory was developed, different energy levels of electrons were admitted to be housed in each electronic shell and, further, that two electrons do not share a similar level in the same shell. The positions occupied by electrons in motion were termed **orbital**. The electrons lying in a shell are displaced inside a strip, receiving or distributing energy fractions termed **quantum** (singular) or **quanta** (plural). An electron is in constant motion inside a strip similar to a car moving ahead on a road with one, two, three or more lanes. When the electron gets closer to the atom nucleus, the radius of the circle roamed by it decreases and the centrifugal force increases. Such being the case, electrons occupying more peripheral orbital zones have higher kinetic energy.

By analyzing the band theory of solids from the atom of sodium (Na11), whose distribution of electrons is 2K, 2L6L, 1M, it can be seen that both K and L shells are fully complete, that is, each has the maximum possible number of electrons. M shell has only one electron. According to the classic atomic theory, however, it could house up to eight electrons. In accordance with the current conception, the distributions of electrons in the shells and their energy levels have specific designations.

Each electron or atomic shell has a main quantum number (n) ranging from 1 to 7, designated as follows: n = 1 (K), n = 2 (L), n = 3 (M), n = 4 (N), n = 5 (0), n = 6 (P) and n = 7 (Q).

The secondary quantum number (l) or azimuthal ranges from 0 to 3 with the following corresponding letters: l = 0 (s), l = 1 (p), l = 2 (d) and l = 3 (f). In the diagram of energy levels, each column is designated by a letter **s**, **p**, **d**, and **f** in increasing order. Energy can be built up by electrons in each electronic shell and then occupy a position far more distant or away from the nucleus. Occupying a position at a certain point in the orbital zone is a probability. The diagram of energy level shows that an electron is never stationary and can be found in any orbital zone of its shell, which highlights the current electronic sphere pattern.

The atomic model by Rutherford-Bohr and Thomson illustrated below, pinpoints the energy levels of electrons:

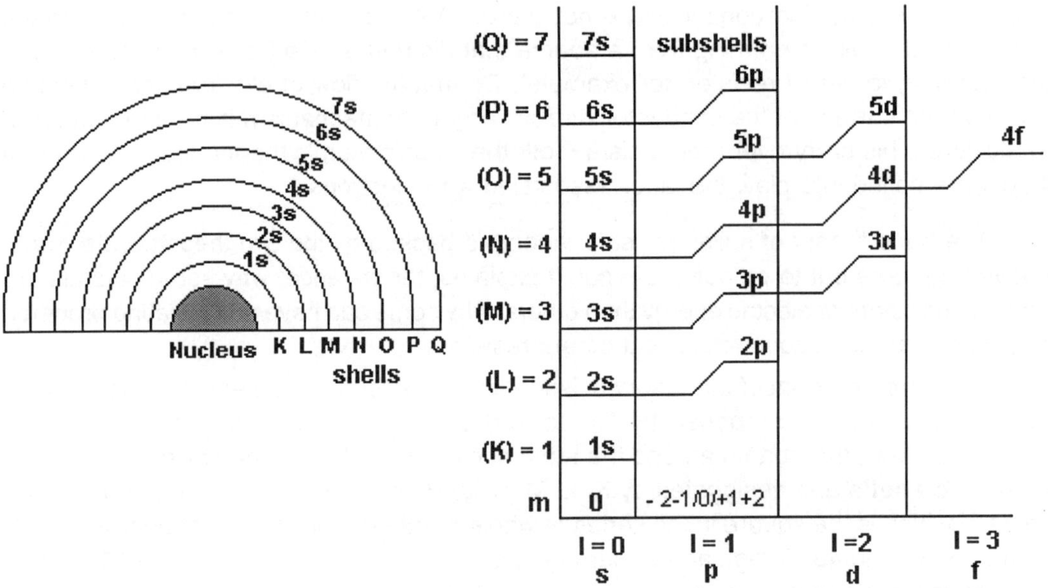

Fig. 73. – **Left**: Illustration of the atom model by Rutherford-Bohr and Thomson showing the seven electron shells (from K shell to Q shell). **Right**: Diagram of energy levels of electrons. The main (**n**) and secondary energy levels (**l**) are represented by the letters s, p, d, f, where **n** is the main quantum number and **l** the secondary quantum number. The levels shown in the example 2p, 3p, 4d and 4f are all subshells of electrons energy.

The letter m or Ml represents the magnetic quantum number. In a certain subshell, the magnetic quantum number in the central orbital equals zero; right orbits m = +1, +2, +3; left orbits m = –3, –2, –1. An orbit can, at most, house up two electrons revolving on their own axis (spin) or in opposite directions (first assertion of Pauli exclusion principle). The spin's quantum number (whose values are –½ and +½) is represented by the letter **s** or **Ms**. By using Pauling's diagram, one can draw up the energy level diagram for orbital electrons with values n, l, m (Ml) and s (Ms).

In the atom of sodium (Na11) taken as a pattern for analysis, three shells are available with the following distribution of electrons: $1s^2/2s^2 2p^6/3s^1$, in other words, two electrons in the 1s subshell of the K shell, two in the 2s subshell, six in the 2p subshell of the L shell and one in the 3s subshell of the M shell.

Atoms of sodium placed side-by-side cause a number of small strips to arise; if juxtaposed, they form the band, the width of which is dependent on the distance **r** between the atoms.

A valence electron forms a narrow strip, but the juxtaposition in the strips produced by a large number of electrons results in a band. The number of narrow strips inside band 3s is far and far greater, as larger numbers of sodium atoms are taken into account, a condition in which electrons can most likely be found. Thus, the energy produced by each valence electron from each atom of sodium, when juxtaposed, fails to form lines and forms a band instead, in the example given, the valence band. The same happens to electrons roaming internally or lower orbits, making up bands 2p, 2s and 1s. Each band has an individual energy level, the higher the farther from the atomic nucleus.

In the case at issue, where the atom of sodium is taken as an analysis model, the electrons that lie up to level 2p, levels 1s and 2s included, are all "armored"; in other words, they are not set in motion even if an electric load is applied. According to the new deuterotomic theory being proposed, the internal structure–where the "armored" electrons are encountered–is that of neon (Ne10) noble gas, whose last shell (L shell) of electrons is completed and neutralized by series-equivalent bindings. The said noble gas that has become electrically neutral and gone through the physical volume-reduction phenomenon also became the genetic pole of sodium deuterotome. For that reason, the internal electrons of noble gas deuterotomes are "armored". The valence electron of sodium (Na11) is linked to the deuterium D11 of that deuterotome. Should potassium be used as the analysis model, the valence band would then be 4s1; the internal bands would be part of noble gas argon (Ar18), the energy levels of which are (Ne) $3s^2$, $3p^6$ plus the aforesaid 4s1. Similarly, electrons that take part in argon's structure are "armored" because that noble gas is the genetic pole of potassium deuterotome, where D19 is the only post-noble gas deuterium.

The energy level for electrons lying in each orbit is inversely proportional to the square radius of the electron's (n) orbit, the value of which is a series of whole numbers from 1 to 7. That energy can be calculated, by means of the following equation:

$$En = \frac{-13.6\,eV}{n^2}$$

where **En** is the electron's energy at a certain orbit, **n** is the orbit's number pursuant to the electron in consideration while the negative value –13,6 eV is the energy that corresponds to the electron's ground state in K shell or n = 1. For example, if n = 1, by replacing En equation: $En = -13{,}6\ eV/1^2$ where En = – 13,6 eV. Also, if n = 3, by replacing En equation: $En = -13{,}6\ eV/3^2$ where En = – 1,5 eV. The Table below shows the En value in each orbit:

n = 1	En1	=	– 13.60 eV
n = 2	En2	=	– 3.40 eV
n = 3	En3	=	– 1.51 eV
n = 4	En4	=	– 0.85 eV
n = 5	En5	=	– 0.54 eV
n = 6	En6	=	– 0.37 eV
n = 7	En7	=	– 0.27 eV
n = ∞	En∞	=	0 eV

The values above show that when an electron shifts from an state of excitation n = 7 to n = 6, for example, it loses more energy; if it previously had a negative energy of – 0,27 eV, the new value will be – 0,37 eV, its excitation energy being reduced. The same thing happens to an electron shifting from n = 6 to n = 5, from n = 5 to n = 4 and so forth, until n = 1 is reached, a condition in which En value is –13,6 eV. If the electron lies at an infinite distance (n = ∞) from the proton, En value will then be equal to zero eV (0 eV). Consequently, the electron that gets more energy than En will express that excess as kinetic energy or excitation energy.

An electron lying at n = 1 orbit, possesses an energy level – 13,6 eV. That energetic value characterizes the electron in ground state. **Excitation energy** is defined as an excitation state **n** for the energy level above the ground state which can be calculated by means of the equation **Ee = En – E1** where **Ee** is the excitation energy of an electron lying at a certain orbit; **En** is the energy level in the orbit in consideration; and E1 is the electron's energy level at n = 1 orbit. This equation makes it possible to calculate an electron's excitation energy in each orbit.

For orbit n = 1 we have: Ee1 = En1– E1. If we replace the equation figures by their respective values (as shown in the Table), we then have: Ee1 = – 13,6 eV – (– 13,6 eV) where Ee1 = – 13,6 eV + 13,6 eV and Ee1 = 0 eV. This shows that an electron at n = 1 orbit (K shell) is in ground state.

The calculation of Ee2 for an electron at n = 2 orbit is as follows: **Ee2 = En2 – E1**; if we replace the equation figures by their respective values: Ee2 = – 3,4 eV – (– 13,6 eV) where Ee2 = – 3,4 eV + 13,6 eV and Ee2 = 10,2 eV. This shows that an electron at n = 2 orbit (L shell) is in excited state. Thus, electrons lying in other orbits are also in excitation state, with higher energy levels. The calculations for all other orbits produce the following values: n = 3 (M shell) equal to 12,1 eV; n = 4 (N shell) equal to 12,8 eV; n = 5 (0 shell) equal to 13,05 eV; n = 6 (P shell) equal 13,22 eV and n = 7 (Q shell) equal to 13,32 eV, thus demonstrating that all the states above n = 1 are excited.

The energy levels of electrons can be represented in the diagram, but the Table pinpoints their locations at once. A list of some elements with their shells and subshells of energy is given below. The subshells of noble gases will be referred to right at the beginning and the following ones will be shown in parentheses ( ) which will contain their own individual series. Each noble gas is a unit provided with "armored" orbital electrons; in other words, not influenced by external electric loads.

H1 $1s^1$	K19(Ar)$4s^1$	Y39(Kr)$5s^2$–$4d^1$
He2 $1s^2$	Ca20(Ar)$4s^2$	Mo42(Kr)$5s^1$–$4d^5$

	Sc21(Ar)$4s^2$–$3d^1$	Ag47(Kr)$5s^1$–$4d^{10}$
Li3(He)$2s^1$	Ti22(Ar)$4s^2$–$3d^2$	In49(Kr)$5s^1$–$4d^{10}$–$5p^1$
Be4(He)$2s^2$	V23(Ar)$4s^2$–$3d^3$	**Xe54(Kr)$5s^2$–$4d^{10}$–$5p^6$**
B5(He)$2s^2$–$2p^1$	Cr24(Ne)$4s^1$–$3d^5$	Cs55(Xe)$6s^1$
C6(He)$2s^2$–$2p^2$	Mn25(Ar)$4s^2$–$3d^5$	La57(Xe)$6s^2$–$5d^1$
N7(He)$2s^2$–$2p^3$	Fe26(Ar)$4s^2$–$3d^6$	Ce58(Xe)$6s^2$–$5d^1$–$4f^1$
O8(He)$2s^2$–$2p^4$	Co27(Ar)$4s^2$–$3d^7$	Pr59(Xe)$6s^2$–$4f^3$
F9(He)$2s^2$–$2p^5$	Ni28(Ar)$4s^2$–$3d^8$	Gd64(Xe)$6s^2$–$5d^1$–$4f^7$
Ne10(He)$2s^2$–$2p^6$	Cu29(Ar)$4s^1$–$3d^{10}$	Dy66(Xe)$6s^2$–$4f^{10}$
Na11(Ne)$3s^1$	Zn30(Ar)$4s^2$–$3d^{10}$	Yb70(Xe)$6s^2$–$4p^{14}$
Mg12(Ne)$3s^2$	Ga31(Ar)$4s^2$–$3d^{10}$–$4p^1$	Lu71(Xe)$6s^2$–$5d^1$–$4f^{14}$
Al13(Ne)$3s^2$–$3p^1$	Ge32(Ar)$4s^2$–$3d^{10}$–$4p^2$	Re75(Xe)$6s^2$–$5d^5$–$4f^{14}$
Si14(Ne)$3s^2$–$3p^2$	As33(Ar)$4s^2$–$3d^{10}$–$4p^3$	Au79(Xe)$6s^1$–$5d^{10}$–$4f^{14}$
P15(Ne)$3s^2$–$3p^3$	Se34(Ar)$4s^2$–$3d^{10}$–$4p^4$	Hg80(Xe)$6s^2$–$5d^{10}$–$4f^{14}$
S16(Ne)$3s^2$–$3p^4$	Br35(Ar)$4s^2$–$3d^{10}$–$4p^5$	Tl81(Xe)$6s^2$+$5d^{10}$–$4f^{14}$–$6p^1$
Cl17(Ne)$3s^2$–$3p^5$	**Kr36(Ar)$4s^2$–$3d^{10}$–$4p^6$**	**Rn86(Xe)$6s^2$–$5d^{10}$–$4f^{14}$–$6p^6$**
Ar18(Ne)$3s^2$–$3p^6$	Rb37(Kr)$5s^1$	

Fig. 74 - Table of energy subshells in some elements.

The band theory explains how electrons move about in solids such as crystals, pure metals, metal alloys and ceramics, as well as in the transportation and blockade mechanisms in conductors and the insulation of an electric current, respectively. Electrons and photons participate in both electrical and thermal conductivity.

Orbital electrons in their shells and subshells energy levels can be outlined, based on the AB space that represents the shells in the atom of sodium (or deuterotome), in accordance with Rutherford-Bohr and Thompson model.

The energy levels, from the nucleus to the lower strip of 3s level are termed **occupied states**, for the number of electrons is complete in the 2N and 2N + 6N bands. Beyond the lower strip of 3s level, **empty states** (unoccupied) are encountered, where electrons can roam about. In the example given, the 3s band is termed **valence band**, where the valence electron of sodium (Na11) can be found.

When an electric load is applied on the material, with the valence band half full, the electron acquires energy and is moved toward an empty band, placed above the Energy gap band (Eg), which lies in the 3p strip, termed **conduction band**, since this is where the transportation of the **electron** of electric current occurs. Thus, the place from where the electron of the valence band came, then takes on a positive load (a part of sodium ion that lost an electron). This place is termed **hole** to indicate the absence of electrons. The electrons carry-over system is shown in the figure below:

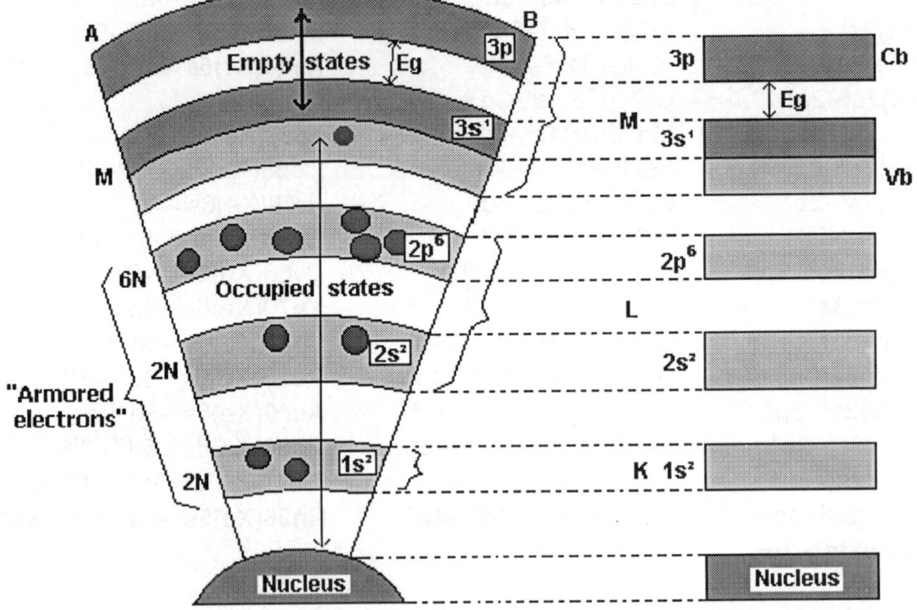

Fig. 75 – Energy levels of electronic subshells in 1s, 2s, 2p and 3s situated inside K, L and M electronic shells. L shell has a 2p subshell. Around the atom are the valence (Vb) and conduction (Cb) bands intermediated by a gap called **energy gap** (Eg). The level of Fermi energy (E_F) lies between the occupied states and the empty states. Electrons in the K shell and L shell are called "armored electrons."

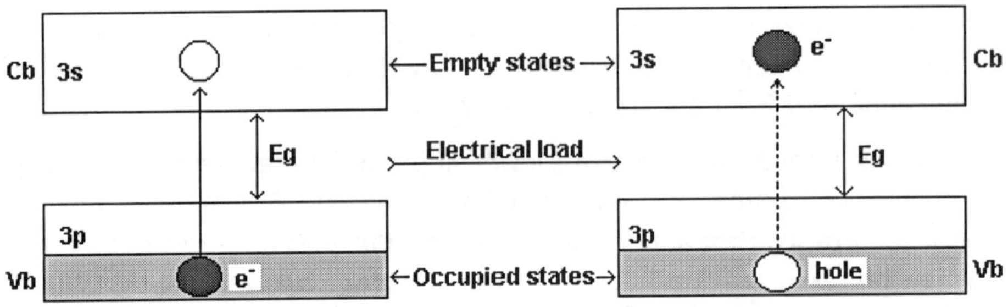

Fig. 76 - **Left**: shows the valence band (Vb) whose lower strip is occupied by the electron (e^-) of sodium (Na11) valence; this electron lies at 3s energy level. The line that separates the white strip from the grayish one in 3s corresponds to Fermi energy level. **Right**: electron (e^-) left the strip of occupied states, went past the energy gap (Eg) strip and reached the conduction band (Cb) where empty states are encountered when an electric load (110 V, for example) is applied on the conductor metal. The place from where the electron left is called **hole**. Sodium without its orbital electron now became sodium ion (Na+). At the temperature of 0 K (zero Kelvin), all the electronic states are occupied below Fermi energy level. The space between the two bands (Vb and Cb) is called gap (interval). That interval is provided with Energy gap (Eg). The value of that energy is important to electrical conductivity: it is lower in semiconductors (around 1.1 eV) and higher in insulating or dielectric ones, around 5 and 8 eV.

The figure below illustrates sodium (Na11) deuterotome composed of the genetic pole (neon) and one deuterium (D11). The electron (e^-) lies inside the valence band. This electron participates in the carriage of the electrons flow supplied by the source. Electrons in 1s and 2s + 2p bands are a part of the neon structure that makes up the genetic pole and do not participate in the carriage of electric current, being therefore, called "armored electrons", since they are not susceptible to the action of an electric current. It should be brought to mind, however, that all those electrons that belong to neon's electronic shell, in the example given, are not "armored" under any conditions. During the electroconversion process, for example, the electron that lies in the K shell (1s subshell) can be captured by a proton situated in the nucleus; Auger electron is one placed in any electronic shell that can be ejected outwards by an extranuclear stimulus.

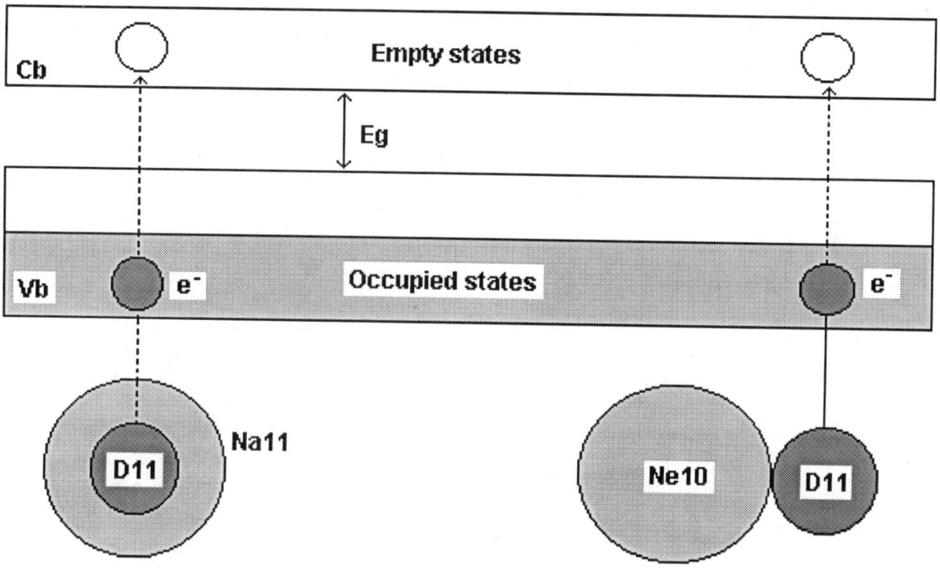

Fig. 77 - Illustration of sodium (Na11) deuterotome with the genetic pole composed of noble gas neon (Ne10) and the only post-noble gas deuterium with its valence electron (top view). **Left**: the electron is in the valence band (Vb) below Fermi energy level (E_F) where the electronic states are all occupied; above the Fermi level the states are all empty. Should sodium deuterotome be stimulated by an external electric load, the electron will move from Vb to Cb, going past the strip of Energy gap (Eg). The place where the electron (Vb) originated is empty and called hole. **Right**: illustration of the same sodium deuterotome (side view) where the genetic pole can be seen as well as the only post-noble gas deuterium (D11) with its orbital electron.

Photons provided with an energy level higher than that of Eg are strongly absorbed resulting in pairs of **electron-hole**. This process makes it possible to use semiconductors in detectors of electromagnetic radiations.

The opposite process, where photons with energy level higher than that of Eg are strongly emitted, in the combination of pairs of electron-hole, is termed "luminescence."

This process is the operation base of the emitting diodes of light and lasers (light amplification by stimulated emission of radiation semiconductors).

Energy gap is the one contained in the strip that separates the valence band (Vb) of the conduction band (Cb) in a solid semiconductor. In an insulating crystal, at the temperature of T = 0 K, the last band from the nucleus of the atom to the periphery, termed **valence band,** is completely full. When the temperature is above zero (T > 0 K), the electrons in the valence band can acquire enough thermal energy to reach the next, more peripheral band, termed **conduction band,** which was empty at zero Kelvin (T = 0 K). Transference of electrons to the conduction band leaves, in the valence band, a state that acts as bearers of positive electric load, termed **holes** (or **lacunas** for some authors). Thus, a hole is the place where an electron (e^-) once existed in the valence band, which migrated to the conduction band; the empty place or hole has the equivalence of a positive electric load (+). Should an electric current or external electric field be applied on the semiconductor, the electrons in the conduction band as well as the holes in the valence band will then produce an electric current or flow of electrons.

For industrial purposes, there are a number of semiconductors of pure metal (silicium and germanium), or forming metal alloys (gallium nitride, GaN; cadmium sulfide, CdS; gallium phosphide, GaP; cadmium telluride, CdTe; Indian arsenide, InAs; Indian phosphide, InP; gallium sulfide, GaS; aluminum arsenide, AlAs), whose high energy gap (Eg) values, expressed in eV, correspond to the largest wavelengths (expressed in μm). The table below shows the values for each semiconductor or insulator:

Insulator /Semiconductor	Eg (eV)	λ (μm)
SiO_2	8.0	–
AlN	5.90	0.21
C	5.0	–
GaN	3.40	0.36
Cd	2.53	0.49
GaP	2.26	0.55
AlAs	2.16	0.57
CdTe	1.50	0.83
GaAs	1.43	0.86
InP	1.35	0.92
Si	1.12	1.11
Ge	0.67	1.88
InAs	0.35	3.54
InSb	0.18	6.87

Fig. 78 – Table of pure metals and metal alloys that function as insulators (**I**) and semiconductors (**S**). As the energy gap (Eg) of metals and alloys rise above 1.5 eV, they become insulators; below 1.43 eV they function as semiconductors. Carbon's Eg is 5 eV whereas silicium oxide's (SiO2) is 8 eV, both insulators; silicium's (Si) Eg is 1.12 eV while germanium's (Ge) is 0.67 eV, both semiconductors.

The processes of absorption and light emission between the valence and conduction bands in dielectrics (insulators) are similar to those of semiconductors. However, as the energy gap (Eg) in dielectrics is to the tune of 10 eV, the photons in the visible area fail to have enough energy so that transitions can be produced between the valence and conduction bands. That's why insulating crystals such as diamond, quartz and NaCl, for example, are entirely transparent to visible light.

Thus, the main difference between a semiconductor and an electric insulator (dielectric) lies in the Energy gap (Eg) value, that is, in the energy contained in the interval or space between the valence and conduction bands. The conductivity of electrons from the valence band to the conduction band is greater the higher the temperature and the lower the energy gap that separates the two bands. One can say that when the Eg decreases, that is, as the valence and conduction bands become far wider and the wavelength (λ) increases, the material then shifts from a condition of electric insulator to that of electric semiconductor. The electrical conductivity in semiconductors is far smaller than that of metals while that of dielectrics is null or minimal.

10.2 – Contac potential among metals

When two plates of different metals with different Fermi energy levels (E_F) are put in contact, a potential difference, called **contact potential** arises between them. The electrons closer to the border of the metal (plate M1 in the scheme to be referred to) with high Fermi energy level, can easily move almost freely to occupy the empty states above the Fermi energy level of the other metal (plate M2 in the scheme). When the two metal plates are put in contact, electrons flow from plate M1 to plate M2, because the Fermi energy level is far smaller in the latter. Thus, plate M1 that had lost electrons, acquires a positive electric load (+) because its deuterotomes were changed into metal ions. The electrons that pull out from the empty states of space W1 to space W2 are removed by an energy Wo termed **work function.** Thus, Wo = W2 – W1.

The electrons in the plates cannot move too far away from the contact borders and are confined in small areas close to the surfaces. One of the plates, the one with lower Fermi energy level, becomes negatively loaded due to the electrons it gets from the other; conversely, the one that gave out electrons becomes positively loaded because its deuterotomes became metal ions, that is, the positive load is obtained by the proton's electric load whose orbital electron is absent. For that reason, a potential difference ΔVe is simply established so as to put in contact the surfaces of the two metals. This potential difference is known as contact potential.

The energy (Wo) necessary to pull out an electron of Fermi energy level to the external face of the metal is called **work function**. The difference in contact potential (ΔV) is determined by the difference between the work functions of the two metals W1 and W2, expressed by the following equation:

$$\Delta V = \frac{W_2 - W_1}{e}$$

where **W1** and **W2** are the work functions in each metal, expressed by eV and **e** is the electron's electric load, expressed in Coulomb (equal to 1.602×10^{-19} C.).

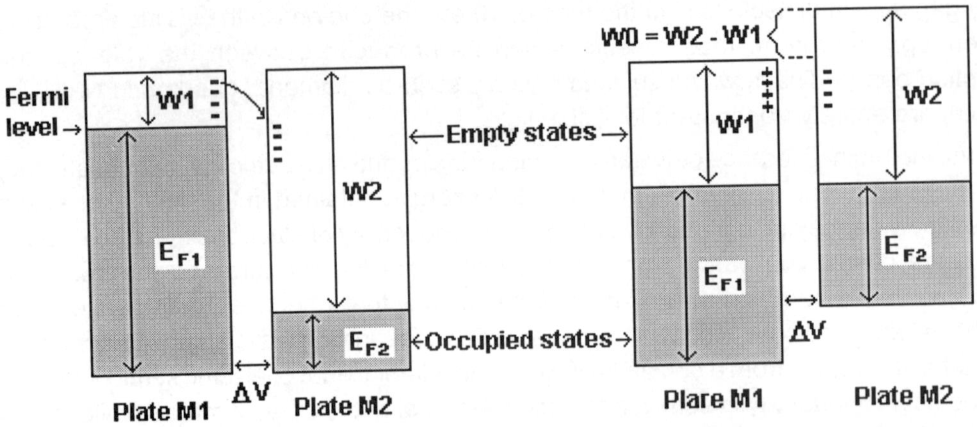

Fig. 79 - **Left**: two plates of different metals M1 and M2 with different Fermi energy levels (E_{F1} and E_{F2}) in contact with each other create a potential difference (ΔV). Wo is the work function and represents the energy needed to remove an electron from plate M1 with a W1 load to plate M2 with a W2 load; the electron pulls out from the interior of the metal of plate M1, crosses the surface and is then captured by plate M2. **Right**: the two plates of M1 and M2 metals placed with their Fermi energy levels juxtaposed. The electrons of M1 are transferred to plate M2 where they will occupy empty states. The exit of electrons from M1 to M2 creates empty states that correspond to positive electric loads (+) and M2 acquires electrons with negative electric load, thus creating a potential difference between them.

10.3 – Possible contact potential between Proton/Neutron (P/N)

When deuterium is formed, the proton (P) and the neutron (N) share a mutual contact surface. The proton's electric load is higher than the neutron's; the electrons that make up the proton load can be housed in the interstice of lentons plates that form the basic structure of both the proton–which makes up light hydrogen–and the neutron. Similarly to two plates of different metals that produce a potential difference or contact potential when put in contact, the proton and the neutron in spontaneous contact–as is the case of deuterium–should produce a potential difference. The contact potential that is an attraction force is associated with the binding energy per nucleon (Ben). In heavy deuterotomes, where "alpha-appendix" exists, alpha units are available with three or more neutrons in contact with each other. A second neutron in contact with a proton/neutron unit will generate another unit made up of proton/neutron/neutron (P/N/N). Therefore, the contact potential between the two neutrons (N/N), in the example given, will be smaller than that existing between the proton and neutron (P/N junction). Also, the binding energy per nucleon (Ben) gets lower as the heavy deuterotome acquires a greater number of neutrons. The contact potential P/N/N is smaller than P/N.

Thus, in comparison with metals, the contact potential can be analyzed between the proton and the neutron (P/N) based on the same data. The two structures (proton and neutron) formed by plates can be shown. The electrons can be housed in the interstice of the proton and neutron between the plates, because the proton mass is 1,836 times greater than the electron's. Accordingly, the proton and the neutron are two plates whose surfaces are in contact. The proton's Fermi energy level is far greater than that of the neutron. This contact enables the proton's electrons to be displaced from the proton to the neutron. The neutron's plate structure is negatively loaded due to the electrons it got whereas that of the proton is positively loaded because electrons were lost. From where the electrons originated, an empty state arises that functions as a positive load. All these electric loads are confined in small areas between the proton and the neutron, thus creating a potential difference or contact potential.

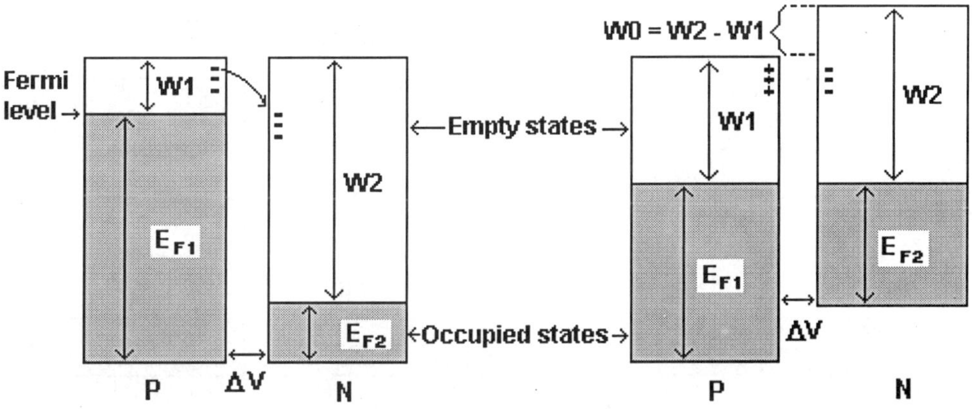

Fig. 80 - Contact potential in the proton/neutron (P/N) junction. **On the left**: illustration of a proton (P) and a neutron (N) whose surfaces are in contact. Fermi energy level is greater in the proton (E_{F1}) than in the neutron (E_{F2}). The electrons, stored in the interstices of the plates that make up the proton, go to the neutron's surface, carrying along a part of the proton with negative electric load. The loss of electrons (or photons) to the proton plate (P) creates an electronic positive state and generates a potential difference (ΔV) or a contact potential between them. **On the right**: the work function of the proton (W1) is the energy needed to remove an electron (or a photon) from the Fermi energy level outside the proton plate. The contact potential between the two plates is determined by the difference between the work functions of the two plates ($Wo = W2 - W1$). The contact potential is given by the equation $\Delta V = W2 - W1/e$, where **e** is the electron's electric load or the photon.

10.4 – Simplified microscopic model of electrical conductivity

Sodium (Na11) deuterotome that possesses one single valence electron in the last shell (monovalent) can be adopted as a model of metal for the analysis of electromagnetic waves and electrons carriage in a one-dimensional grid system. When the electron on the peripheral orbital zone 3s, leaves the stationary state and sets about the vibration state, induced by an external electric field, it creates an electromagnetic wave, whose frequency and amplitude depend on the physical nature of the conductor material, on the intensity of the external electric field and the temperature of the conductor wire.

Should the stimulus that generated the wave stimulate **n** deuterotomes in the same frequency, it will also generate **n** waves continuously. A simplified and one-dimensional scheme may illustrate a condition that will make it possible to analyze what occurs at the beginning of the physical phenomenon of electrical conductivity. A situation close to reality would be a three-dimensional scheme, albeit extremely complex. Two sodium (Na11) deuterotomes, each with their valence electron represented on the valence band (Vb) are available. The said electron moves towards the conduction band (Cb), crossing the Energy gap (Eg) strip. Where the valence electron left from, an empty space or hole ensues.

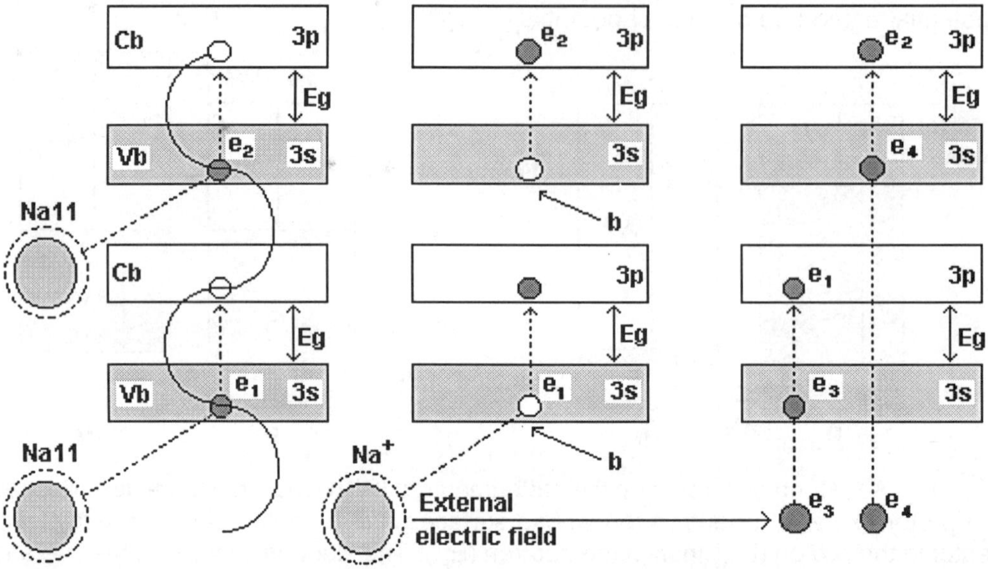

Fig. 81 – Simplified one-dimensional microscopic model of electrical conductivity. **I** - Two sodium deuterotomes with their respective orbital electrons in the valence band (Vb). Electromagnetic waves formed by vibrations from electrons caused by an external electric field. **II** – Electron e_1 moves from valence band (Vb) in 3s toward that of conductivity (Cb) in 3p; the place from where electron e_1 left, an empty or hole (b) arises which represents a positive load because element sodium (Na11) became a metal ion (Na+). The same phenomenon occurs in electron e_2. **III** – The external electric field supplies the electron flow (e_3 and e_4) and each will occupy a hole. The electron that is occupying a conductor band shifts from a hole to the band of next valence. That mechanism of electrons carriage (electric current) is repeated continuously for as long as the external electric field is in operation. In superconductor metals or metal alloys, electrons flow freely without the necessity of being transported by the pair of electron-hole.

The stimulus from the external electric field fuels up the mechanism that causes Vb to give away electrons and Cb to receive them, alternately. The electron in the valence band (Vb) needs to cross the Energy gap (Eg) strip to reach the conductivity band (Cb). If Eg is between 5 to 10 eV, the electron cannot get through and the material falls into the category of insulator to an electric current or a flow of electrons.

The vibration from electromagnetic waves caused by the stimulus of an external electric field can vary in **length**, **frequency** and **amplitude**. The table in fig. 78 shows that Eg is higher in metals that can emit electromagnetic waves with shorter lengths. Thus, aluminum nitride (AlN) has an Eg 5.9 eV and wavelength 0.21 μm, while Indian antimonide (InSb) has a low value of eV (0.18 eV) and a long wavelength (6.87 ìm). For the sake of analysis, the basic elements of each wave is highlighted below:

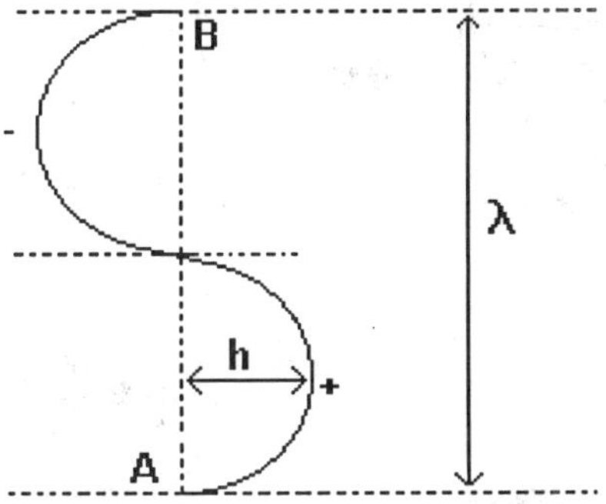

where lambda (λ) is the **wavelength** that goes from A to B; **h** is the **amplitude** of the wave; (+) the **positive band**; (–) the **negative band**. **Frequency** (f) is the number of waves formed in a unit of time. Thus, **f = n/t,** where **f** is the frequency, **n** the number of waves and **t** the time of one second. One wave per second corresponds to the frequency of one Hertz.

The two demonstrative schemes show the thickness variations in the two bands (Vb and Cb) and the strip of Eg related with the length of the electromagnetic waves.

There are no electronic states in the areas situated between the two bands. They are called forbidden strips or **forbidden gap**, because the electrons are not orbiting the strips, they just roam past them.

Both the electrical and thermal conductivities should be considered in metals. Nevertheless, conductivity has to be analyzed as occurred through a gas. As the electrons in a metal also contribute to thermal conductivity, this can be represented by the equation of classic kinetic theory for a given gas:

$$K = \frac{C \cdot v \cdot l}{3}$$

where **K** (kappa) is the thermal conductivity; **C** the heat capacity by a volume unit; v the molecular speed; **l** the distance between collisions. After being deducted and by replacing the molar **heat capacity,** the equation can be summarized as follows.

$$K = \frac{\pi^2}{3} \cdot \frac{k^2 \cdot n \cdot l \cdot T}{m \cdot v}$$

where **k** is the number of waves in electrons; **l** the distance between collisions; **n** the number of electrons by volume unit; **T** the temperature in Kelvin (K); **m** the mass of an electron (= $9.10953.10^{-31}$ kg); **v** the molecular speed.

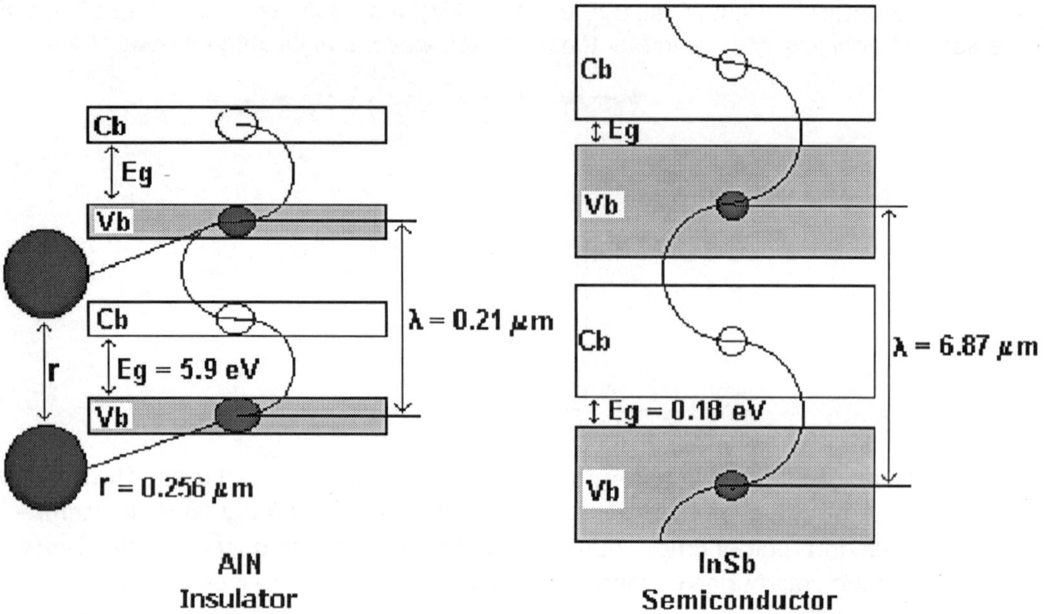

Fig. 82 – Simplified scheme of the two bands (Vb and Cb) and the strip (Eg). **Left**: small wavelength of Vb and Cb with small thickness and great Eg strip. **Right**: great wavelengths of Vb and Cb bands and narrow Eg strip. High Eg (5.9 eV) and small wavelengths (0.21 µm) are values common to insulating materials. The opposite, small Eg (0.18 eV) and long wavelength (6.87 µm) are values common to semiconductors.

The electrical conductivity can be calculated in accordance with the following equation:

$$\sigma = \frac{n \cdot e^2}{m} \cdot \tau$$

where σ (small sigma) is the electrical conductivity; **n** the number of electrons per volume unit; **e** the electron's electric load (= 1.602×10^{-19} C) whose value is a constant; τ (small tau) the time between collisions; **m** the electron's mass (= 9.10953×10^{-31} kg), the time τ (tau) between collisions being expressed in terms of motion speed of electrons (**e**); and (**v**) the molecular speed; hence: τ **= l/v** and by replacing in the previous equation:

$$\sigma = \frac{n \cdot e^2}{m} \cdot \frac{l}{v} \quad \text{or} \quad \sigma = \frac{n \cdot e^2 \cdot l}{m \cdot v}$$

Thus, the electrical conductivity can be calculated in accordance with this reduced equation. However, **l** (distance between collisions) is a parameter hard to be calculated. In the equation for thermal conductivity, that greatness takes on the same aspect. To overcome this problem, the relationship between the thermal conductivity (K) and the electrical conductivity (σ) is used instead. Thus, we have:

$$\frac{K}{\sigma}=\frac{\frac{\pi^{2}}{3}\cdot\frac{k^{2}\cdot n\cdot l\cdot T}{m\cdot v}}{\frac{n\cdot e^{2}\cdot l}{m\cdot v}}$$

After being reduced, the equation that provides the relationship between the two conductivities can be summarized as follows:

$$\frac{K}{\sigma}=\frac{\pi^{2}\cdot k^{2}}{3\cdot e^{2}}\cdot T \quad \text{or} \quad \frac{K}{\sigma}=N_L\cdot T$$

where **NL** is Lorenz **number** and **T** the temperature in Kelvin (K).

Thus, the relationship between the thermal and electrical conductivities depends exclusively on the temperature and would be identical for all metals at a given temperature. This is known as **Wiedsmann-Franz Law**. Also, the constant $\pi^2.k^2/3.e^2$ is known as Lorenz number with an average theoretical value $2.44.10^{-8}$ W.Ω / K² because small variations are produced among metals at 0°C and 100°C. Zinc, for example, whose Lorenz number is 2.31 x 10^{-8} W.Ω / K² at 0°C, at 100°C only 2.33 is reached. Tin (Sn50) number is a bit higher at 0°C (2.52) and 2.49 at 100°C, hence, lower at 0°C.

10.5–Conductor, semiconductor and insulating materials. Superconductors

Conductor materials, also called metals, are those whose valence band (Vb) is half full. This happens when the amount of valence electrons is an odd number. Falling into this category, are the alkaline metals Li3, Na11, K19, Rb37 and Cs55 (all monovalent) plus the noble metals Cu29, Ag47 and Au79, all monovalent and with other valences (copper's valence is 1^+ and 2^+; gold's 1^+ and 3^+) all of them having an odd number of electrons. An electric field applied on these metals produces a flow of electrons, thus featuring an electric current.

The alkaline-earthy metals (Be4, Mg12, Ca20, Sr38, Ba56) are bivalent (2^+) positive, that is, there is a pair of electrons in the last shell or chemical pole, according to the new theory being proposed. The distribution of bands is more complex. Band Vb, which would usually be the last full band, produces a maximum energy level well above the minimum of band Cb. Electrons tend to occupy those states where lower energy level can be found; electrons lying in the highest level of energy of band Vb shift to the lowest level of band Cb and are, therefore, both incomplete. The displacement of electrons occurs because the bands are partially overlapped and the energy gap (Eg) strip is not available. The Fermi energy level (E_F) is well above the occupied states. The following scheme shows the characteristics of bands in metals (on the left) and in semimetals (on the right).

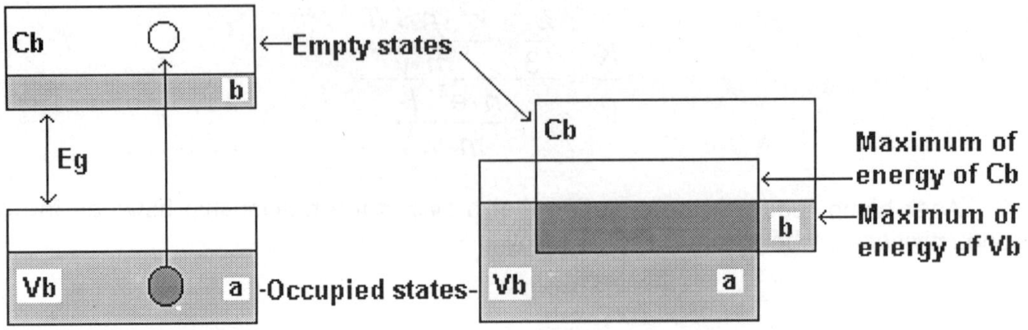

Fig. 83 - **Left**: illustration of the valence (Vb) and conduction (Cb) bands in a metal electric conductor, monovalent (Na11 or K19) positive, with an energy gap (Eg) between them. Band Vb is half full and band Cb has only occupied part, both incomplete. **Right**: illustration of Vb and Cb bands in a semimetal electric conductor, bivalent positive (Ba56 or Mg12). There is no energy gap (Eg) strip; **a** and **b** are occupied states.

Insulating materials such as the crystals of carbon (diamond) and silicium oxide (SiO_2) have their last band (valence band) completely full at the temperature of 0 K. When the temperature is above zero (T > 0 K) the electrons in the valence band (Vb) can acquire enough thermal energy so as to move on to the next band situated farther from the nucleus than the valence band, called conduction band (Cb), which was empty at the temperature of T = 0 K. When an electron abandons Vb and moves on to Cb, an empty state called hole remains. Upon the electron's exit (negative load), the deuterotome functions as a positive ion as follows: Na ——————> $Na^+ + e^-$

A certain insulating material at T = 0 K, when at room temperature (27ºC or 300 K), causes electrons to be displaced from the valence band to the conduction band, called semiconductor. If an electric load is continuously applied on that material, a flow of electrons is carried out by the electron-hole carriage system. The electrons that come from the generating source occupy the hole and undo the metal's ion condition. Soon after this, however, they move on to the conduction band; the electron in this band moves forward by the same mechanism. The material conductivity depends on the number of electrons that migrate to the conduction band; this mechanism is as more intense the higher the temperature and the smaller the energy gap that separates the two bands (Vb and Cb). Low Eg values are found in semiconductor materials (for example, Eg = 1.12 eV for silicium) and high in insulator materials (for example, Eg = 8 eV for SiO2).

The thermal conductivity in metals is two or more times higher than the electrical conductivity. The **electric resistance** of a metal is given by the following equation:

$$R = \rho \cdot \frac{L}{A}$$

where ρ (rho) is the electrical resistivity expressed in W.cm; **L** the length of the conductor wire in cm; A the area of the conductor wire in cm²; **R** the resistance of the conductor wire in Ω (ohm).

Every metal that is a good conductor has a low electrical resistivity. Examples: ρ of copper is 1.7 x 10^{-6} Ω.cm (low); ρ of sulfur is 2 x 10^{17}Ω.cm (very high).

10.6 – Superconductivity. Superconductor materials. Meissner effect.

At a low temperature, the metal resistivity is nearly close to becoming constant. In Fig. 84 on the left, line **b** is almost parallel to line **T**, where **b** is the resistivity and **T** the temperature in Kelvin. That is the feature in an ordinary conductor. The chart on the right points out the resistivity in a superconductor below the critical temperature (**Tc**) when it becomes null or, otherwise, when the resistivity fades away.

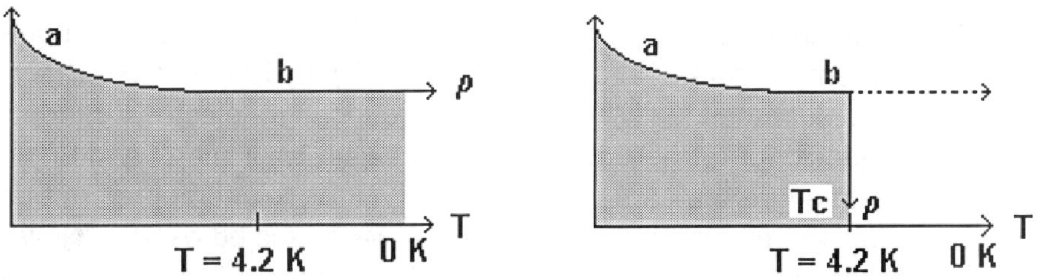

Fig. 84 - **On the left**: in an ordinary conductor, the resistivity in curve **a** drops down to T = 4.2 K; line **b** becomes almost parallel to lineT, that is, the resistivity gets close to the constant at T = 0 K. **On the right**: curve **a** declining down to T = 4.2 K and dropping suddenly to zero, meaning that the resistivity has become infinite because the superconductor resistance to the flow of the electric current has reached zero.

When a metal or a metal alloy has the special property of superconductivity, the resistivity drops down to zero at a temperature T above 0 K (T > 0 K or > –273°C). This temperature is called **critical temperature** (**Tc**). The chart below points out that characteristic. Resistivity rho (ρ) in Tc is not merely very low. In fact, it disappears. Such materials, called **superconductors**, can carry on electric current even in the absence of an electric field or electric load. Electrons in motion interact with the crystal network of superconductors, most likely for causing vibrations. Through the interaction of energy, the electron's energy can be transferred on to a second one. Thus, electrons in motion interact only with other electrons, not with the crystal network. For that reason, the electrical conductivity is infinite, regardless of whether the crystal network is in pristine order or flawed with impurity. The conductivity of electrons is not performed by a system of bands, as happens in semiconductors. In superconductors, electrons move along in sets of two, and no energy is lost, for they fail to interact with the crystal network. The critical temperature (Tc) of superconductivity is determined by the binding energy in the pair of electrons, namely **pair of Cooper**. It should be noticed that the electrons forming the pair are not merely physically connected; they can be set wide apart by a great distance and move away in opposite directions. Some examples of Tc values, expressed in Kelvin (K), in ordinary metals, metal alloys and ceramic of superconductors are shown below:

Ordinary metals		Metal alloys		Ceramic (copper oxides I)	
Superconductors type I		Superconductors type II		Superconductors type II	
Zn	– 0.9	$Nb_{0.3}Ti_{0.7}$	– 9.2	$Yba_2Cu_30_7$	– 92
Al	– 1.2	Nb_3Sn_2	– 16.6	$Bi_2Ca_2Sr_2Cu_3O_{10}$	–110
In	– 3.4	Nb_3Al	– 17.5	$Tl_2Ca_2Ba_2Cu_3O_{10}$	–125
Sn	– 3.7	Nb_3Sn	– 18.1		
Hg	– 4.2	Nb_3Ge	– 23.2		
Pb	– 7.2				
Nb	– 9.1				

In 1909, Kamerlingh Onnes succeeded in liquefying helium (He2/4), a phenomenon that occurred at a temperature of about 4.2K (–268.8°C). In 1911 this physical researcher himself, discovered superconductivity by using mercury as an electrical conductor. At the temperature of liquefaction of helium (4.2K), the resistivity of mercury dropped to zero. When other experiments were carried out, by using nitrogen's temperature that happens at 77 K, other ordinary metals, alloys and ceramic were studied as well. Nowadays, it is common knowledge that the stable material at the highest critical temperature (Tc) of 125K, is copper oxide I, whose composition is $Tl_2Ca_2Ba_2Cu_3O_{10}$ (thallium-calcium-barium-copper-oxygen).

Ordinary metals produce strong magnetic response at temperatures below Tc and are provided with a perfect diamagnetism at T < Tc. It was observed that a superconductor placed inside a magnetic field expels from its interior the force lines of that magnetic field when the temperature (T) of the superconductor gets below the critical temperature (Tc). This phenomenon is known as **Meissner effect**.

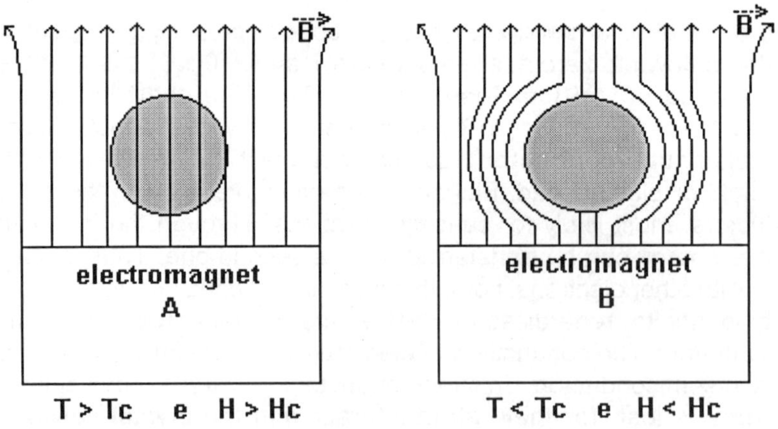

Fig. 85 – Illustration of Meissner effect in superconductors type I (ordinary metals). **A:** scheme of a niobium sphere (Nb) inside a magnetic field B where temperature T is higher than Tc. **B:** magnetic field B is expelled from the interior of the niobium sphere when temperature T is lower than the critical temperature (Tc).

The intensity of a magnetic field **H** influences the appearance of the phenomenon. Thus, the intensity of H has a critical value called **Hc**, above which the material has a normal response at any temperature. So that Meissner effect may occur, **T** must necessarily be lower than **Tc** (T < Tc) and H smaller than Hc (H < Hc). According to the theory in force, magnetization does not originate in atomic magnetic dipoles, but rather in macroscopic currents induced in superconductors with the aid of the magnetic field called **overcurrent**, induced by Faraday effect.

Only ordinary superconductor metals are provided with magnetization throughout interval T < Tc and H < Hc. The critical intensity in the magnetic field **Hc,** however, occurs exclusively in type I superconductors, but two fields Hc (**Hc1** and **Hc2**) are available in type II superconductors. Thus, in type II superconductors one other phenomenon called **state of vortexes** may occur where the magnetic field B is not entirely expelled; a few induction lines remain inside the superconductor. This induction produces bundles of lines inside the metal called **fluxoids**. See next illustration (Fig. 86).

Two theories account for the superconductivity mechanism: the **Theory of London** and **BCS Theory** (formulated by **B**arden, **C**ooper and **S**chrieffer in 1957). The Theory of London is phenomenological, based on classic equations and the BCS Theory is microscopical and entirely quantum, virtually based on the formation of the pair of electrons (pair of Cooper), which have an opposite spin. Thus, the superconductivity state results from the arrangement of conduction electrons, which are aligned two by two to decrease the system's total energy.

The pair of Cooper is characterized by a distance between them called **coherence length** (î), given by the equation î = $\hbar$**.** VF /ð.Ä, where $\hbar$ is the constant of Planck/2ð (= 0.6582 x 10^{-16} eV.s); V_F is the speed of the electrons that form the pair and Ä the energy reduction that the electron undergoes to form the pair of Cooper (the energy level in the pair of Cooper is lower than the energy total in each electron individually). The coherence length (î), expressed in Angstrom (Å), is the greater the smaller is Tc. For example: aluminum (Al13) has a Tc = 1.1 K and î = 16,000 Å; niobium (Nb41) has a Tc = 9.5 K and î = 380 Å. Ceramic whose composition is YBa_2 Cu_3 0_7 has a Tc = 92 K and î = 10 Å.

The critical intensity values of magnetic field **Hc,** expressed in Oersted, have higher values as the **Tc** of the superconductors increases.

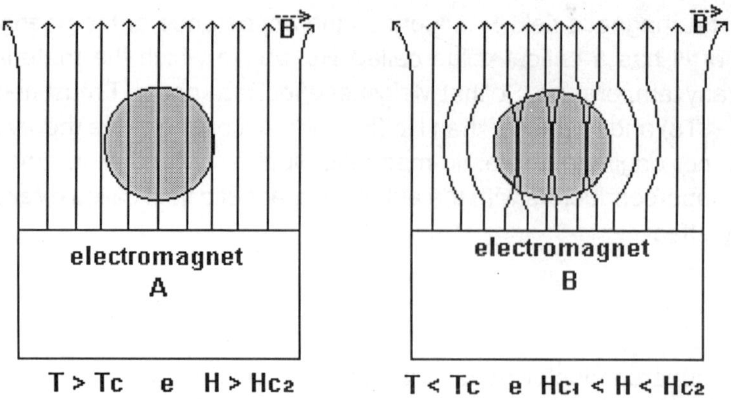

Fig. 86 - **State of vortexes**: the lines that compose fluxoids are called vortexes (= whirl, hurricane). Type II superconductors occur where T > Tc and H > Hc2 the flow of force lines is normal (in A) and where T < Tc and $Hc_1 < H < Hc_2$ the lines of magnetic induction situated more internally go through the superconductor, forming cylindrical bundles called fluxoids and those lying more externally cross from the outside, that is, they are partially expelled from magnetic field B. Vortexes are the filaments that make up fluxoids.

Superconductors are used in coils that bring about intense magnetic fields. The superconductor metal wire performs the winding of coils (spires). Because the wire resistance at a temperature where T < Tc is very low, the wire can be crossed by a high intensity electric current without too much heating being produced. These coils should be kept deep in liquid helium (T = 4.2 K) and are used in electromagnets, in tomography apparatus by magnetic resonance, in motors and special electric tape recorders of great power. Superconductors of high Tc made of ceramics are brittle and inadequate for winding coils. Besides, the critical electric current is not high enough. The necessity of lowering its temperature down to 269°C in order to maintain the superconductance property has hindered and limited their applications.

There are two important junctions formed by an ordinary metal (N), a superconductor (S) and an insulator (I): the NIS junction (ordinary metal-insulator-superconductor) and (2) the SIS junction. These junctions can be represented by plates, as follows:

N	I	S

S	I	S

In the NIS junction if the insulating layer is 100 Å thick, the potential barrier will prevent the electrons from leaving plate N on to S and vice-versa. If the layer is thin between 10 and 20 Å, there will be a huge likelyhood of electrons passing from a plate to the other by means of the tunnel effect; there must necessarily be occupied states in a plate and empty states in the other. A potential **V** applied to one of the plates causes energetic imbalance, the value of which is expressed by eV.

In the SIS junction the two plates are superconductors with a thin insulating layer between them. All happens as if it were a NIS junction, but even if the energy potential V then applied were equal to zero (V = 0 V), an electric current of maximum value exists designated by **Io**, produced by the tunneling in the pair of Cooper electrons. This effect is called **Josephson effect**. The electric current in V = 0 has a quantum nature and can flow towards any of the two directions

CHAPTER 11

Cathode rays, X-rays and Gamma rays

11.1 – Cathode rays

Whenever a high electric load (30,000V, for example) is applied to the cathode of Crookes' tube, the electrons in that pole are thrown against the glass wall that lies ahead. These accelerated electrons can be regarded as electron-projectiles, for they will strike the tube wall. As a result of the collision, the electrons lose kinetic energy and emit continuous X-rays photons, electromagnetic waves also known as Roentgen rays.

11.2 – X rays

The electrons that make up the electronic shells of noble gases (helium, neon or argon, for example) are not used in the mechanism of electrons carriage, that is, the electric conductivity is not dependent on the electrons that take part in the electronic shells that form a noble gas structure. In the diagram that outlines sodium (Na11) energy levels, for example, whose composition in shells is Na11 (Ne) $3s^1$ or Na11 $1s^2\,2s^2\,2p^6\,3s^1$, the electrons that participate in levels $1s^2\,2s^2\,2p^6$ are all armored, for they are not subject to the action of electric loads. However, these electrons are vulnerable during the internal conversion process. A nucleus proton captures an electron in the K shell, called electron K, and a vacant space is then created in that shell. Electron L moves on to K shell, loses energy, and emits characteristic X-ray photon of element (termed **discrete X-ray**).

In short, it can be said that there are two sources of X-rays (accelerated electron and electron in excited state) and two types of X-rays (continuous and discrete).

The accelerated electron or cathode ray produced in Crookes' tube loses kinetic energy upon hitting a target wall and emits a continuous X-ray photon. This process can be symbolically represented as follows:

Electron K (with high kinetic energy) ——> **electron K'** (with low kinetic energy) + continuous X-ray photon

This production process of continuous X-ray is called **bremsstrahlung,** German word for deceleration or breaks, expressing the production of X-ray through the deceleration of electron-projectiles. This type of X-ray is produced outside the deuterotome. This mechanism can be represented as follows:

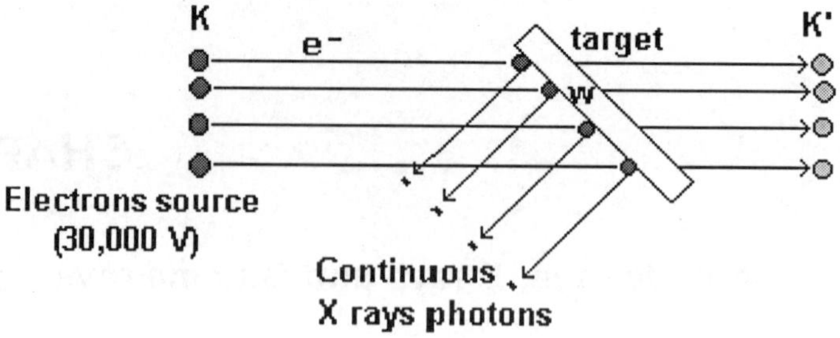

Fig. 87 – Sketch of continuous X-ray production. Electrons-projectiles (cathode rays) are thrown against to target wall (W). The collision against the wall causes the electron's kinetic energy to be reduced and brings about an emission of continuous X-rays photons.

11.2.1 – Continuous X-ray production and bremsstrahlung

Whenever an electric load, an electron, for example, is accelerated or decelerated, it irradiates electromagnetic wave or photons. The energy deriving from that electromagnetic radiation consists of discrete packages of energy or quantum, called **photon**. Thus, a photon is defined as a quantum of electromagnetic energy emitted or absorbed and each radiation photon has energy **E** in accordance with the equation **E = h.v**, where **h** is Planck's constant (= 6.62618 x 10^{-34} J.s.) and **v** the radiation frequency.

A bunch of electrons, which acquired energy **eV** and after being accelerated by an electric discharge of potential **V**, strikes the target, reduces its kinetic energy and takes on a rest state. Electromagnetic waves or photons are emitted as a result of these collisions. During the collisions, the momentum is transferred over to the deuterotome and the electron's kinetic energy drops and photons are then emitted. The kinetic energy of the deuterotome's recoil is low, because the deuterotome is relatively big. Consequently, the electron takes on an initial kinetic energy K and a final lower kinetic energy K'. This deceleration gives to the photon the energy as per the following equation **E = h.v. = K – K'**.

The variation that happens to the electron's kinetic energy results in two distinct conditions, namely: 1) photoelectron production and 2) production of bremsstrahlung radiation.

During the **photoelectron production**, the electron captures a photon, and causes both the acceleration and the energy level to rise, whereas during the bremsstrahlung process the accelerated electron, with kinetic energy K and high energy level, emits a continuous X-ray photon and is changed into an electron of lower energy and lower acceleration (with kinetic energy K'). Schematically, both can be represented as follows:

Photoelectric effect: electron (K) + photon ————>electron (K '), where (K < K')

Bremsstrahlung process: electron (K)——> electron (K ') + photon, where (K > K')

Depending on the electric load applied to the cathode of Crookes' tube, the continuous X-ray wavelength will be longer or shorter. High electric loads (50,000V, for example) produce electromagnetic waves of short length (0.02 nm).

The electronic excitation due to the change in energy level, when an electron shifts from an upper to a lower shell and loses energy and emits X-ray photon characterisitc of element (discrete X-ray), is the basic source for the production of that type of X-ray. The vacancy of one electron in the K shell results from the capture of one electron K by an intranuclear proton. Thus, the internal conversion process begins with an excited nuclear state that can be formed by a previous process that can be either the electron capture or beta decay. The typical X-ray originates in the deuterotome's electronic shells.

11.2.2 – Production of characteristic X-ray of elements (discrete X-ray)

This type of X-ray results from an electron being removed from the electronic K shell of a deuterotome with the lowest energy level in the shell, termed rest energy, the value of which equals –13,6 eV. The process that removes the electron from the K shell is the *electron capture* by the proton of intranuclear location. The vacancy created by the removal of electron K is then occupied by an electron L; the displacement of electron L to the K shell creates a vacancy in the L shell, which will then be occupied by electron M. If M is the last shell, the element may remain as a negative ion, otherwise, a free electron may fill up the shell. The illustration below, sketched in keeping with the atomic model of Rutherford-Bohr and Thomson, shows the electrons in the K, L and M shells.

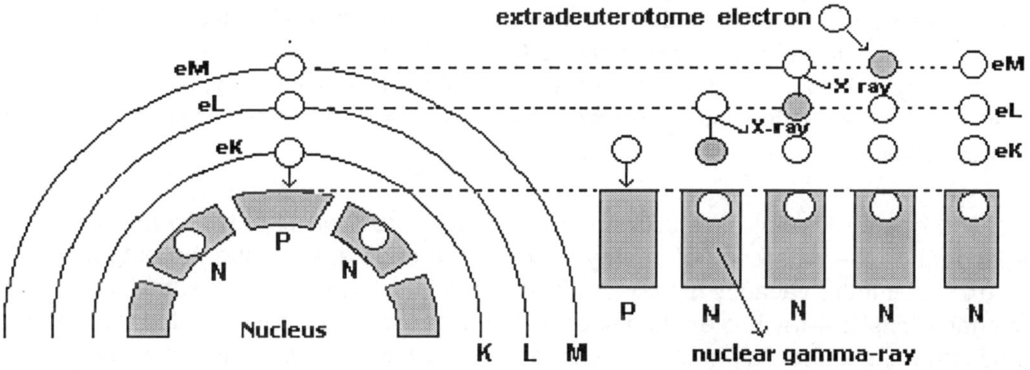

Fig. 88 - Electron eK is captured by the nuclear proton P. A vacancy or hole is available in the K shell. An electron eL will move from L shell (where it possessed energy –3.4 eV) and will be transferred to the K shell (where it will acquire a rest energy –13.6 eV) and emit a photon. The same occurs to electron eM that will leave the M shell (where it has an energy –1.5 eV) and will move on to L shell (where the electron's energy equals –3.4 eV) and emit a photon. These photons emitted are electromagnetic waves, called **discrete X rays.**

Accordingly, when an electron is displaced from its normal configuration by a stimulus (electron capture, for example), the deuterotome experiences a very short period of excitation of one nanosecond or less in many solids. Electronic shells tend to reset the position of their electrons. However, when the electron moves from an upper to a lower shell, it will emit a photon and will lose energy; this energy loss will be far greater as more inwardly are the shells.

The energy released during the transition from the excited electronic state to the one of rest or, conversely, a lower kinetic energy, is a characteristic X-ray photon, whose energy level is given by the energy difference between the initial and the final states.

The electron captured is generally an electron K. The characteristic X-ray of element receives the name of the shell from where the first electron came, where the first vacant space was created and plus the Greek letter (α, β, γ or δ) being added to the shell from where the electron derived to reset the vacancy. The scheme below shows these designations.

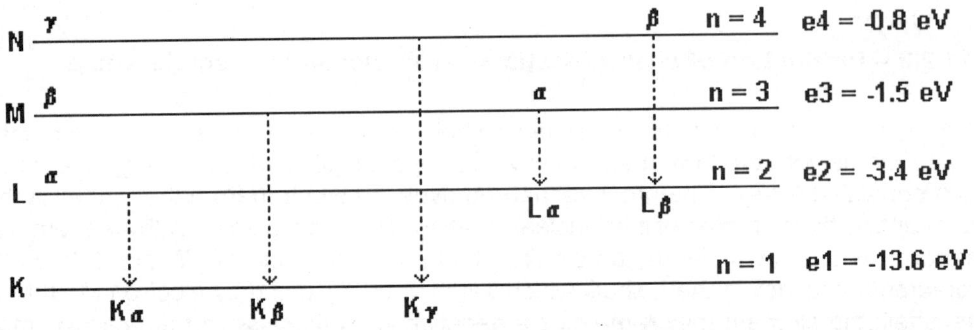

Fig. 89 - Scheme showing the production of characteristic X-ray. If an electron is pulled out of the K shell and the vacancy is filled by an electron coming from the L shell, the wave in K spectrum will be designated Kα and will have a lower energy level; the photon emitted is known as X-ray of K shell. If the electron comes from the M shell, it will produce a wave in the spectrum of higher energy defined as Kβ. In the Kγ wave the energy will be higher. If an electron is initially pulled out of the L shell, the same type of resetting will take place and the photon emitted will be the X-ray of L shell.

The energy loss of an electron when it moves from an upper to a lower shell is dependent on the difference of energy between the original shell and the outgoing one. For example: if the vacancy is temporarily created in the K shell of a deuterotome, then the characteristic X-ray K is emitted soon after the vacancy has been filled. If the electron comes from the L shell, a photon Kα is then produced, the energy of which will be similar to the difference between the one in electrons situated in both the K and L shells. Should the electron that fills up the vacancy have originated from the M shell, photon Kβ is produced with a slightly higher energy and so on. The maximum energy of a photon (of the K series) is obtained when the vacancy is taken up by a free electron emerging from outside. The corresponding photon's energy is then simply provided by the energy connected to the K shell. Vacancies created in the most outlying shells resulting from the displacement of electrons to occupy vacancies in the lowermost or more inward shells are forthwith filled with emissions from characteristic X-rays of L, M series, etc.

Because they have high energy, X-rays in the K series are generally more significant in terms of practicability. Their energy increases regularly when the number of photons in elements rises. This energy is around 1 keV for sodium (Z = 11); for gallium (Z = 31) equal to 10 keV; for radium (Z = 88) equal to 100 keV. For discrete X-rays in L series, the energy fails to reach 1 keV up to Z = 28 and 10 heV up to Z = 74. Because the energy in characteristic X-rays is specific for individual elements, they are frequently used to analyze elements in samples whose composition is unknown.

For a deuterotome in excited state, the ejection of Auger electron turns out to be a competitive process for discrete X-ray photon to be emitted. Fluorescence is a physical phenomenon that results from the emission of this photon during the resetting process of electrons in the shells of a deuterotome in excited state.

A great number of different physical processes can lead straight to the production of excited deuterotomic states from where characteristic X-ray originate. The excitation mechanisms of greater practical importance for laboratorial production of discrete X-ray and continuous X rays sources are: 1) excitation by external radiation (cathode rays or accelerated electrons); and 2) excitation by radioactive decay (beta decay and electron capture).

11.2.3 – Sources of characteristic X-rays of elements

a) Capture of electrons. The capture of an electron K by the nuclear proton creates two excited states, one at the electronic shells level, one other at nuclear level. Therefore, whenever an electron K is captured by the nuclear proton, the orbital electrons experience a state of electronic excitation during the resetting phase in the shells; when an outlying shell moves into one situated more inwardly, the electron loses energy and emits a photon that corresponds to the discrete X-ray, a typical feature of that element.

When the nuclear proton captures electron K, the former is transformed into a neutron and emits a neutrino as per the equation:

$$P + e^- \longrightarrow n + v$$

where **P** is the proton, **e**$^-$ the electron captured, **n** the neutron resulting and **v** the neutrino emitted. The process of electron capture does not occur with the free proton, but rather with the nuclear proton. The general equation of which is as follows:

$$^{A}_{Z}X_N + e^- \longrightarrow {}^{A}_{Z-1}X_{N-1} + v$$

When the proton changes into a neutron through an electron being captured, it creates an excited nuclear state. When the nucleus shifts from that state to one of rest, it will emit one or more photons, corresponding to the **nuclear gamma rays**. The energy in each photon is exactly the difference between the initial energy and the final nuclear state, less the negligible correction due to the recoil energy by the nucleus. That energy varies typically from 100 keV and a few MeV. During this process of electron conversion, the proton changes into a neutron and decreases the Z deuterotomic number by one. The resulting deuterotome still has the same total number of electrons, albeit one less electron in the electronic shells. The process of electron capture creates a vacancy or hole in one of the internal shells. As soon as this vacancy is filled up, the X-ray is generated with spectrum lines typical of producing element.

The decay alters the rest state of the electron in the K shell and also creates an excited nuclear state in the producing deuterotome so that the characteristic X-rays can be likewise accompanied by emissions of gamma rays during the subsequent reduction in the excited nuclear state.

The electron capture by the deuterotome's nuclear proton can bring about another kind of continuous electromagnetic radiation known as inner bremsstrahlung; during that decay process, an orbital electron is captured by the nuclear proton and, for that reason, it should, to some extent, be subject to acceleration. According to the current atomic theory, it can be inferred that an accelerated electric load must emit electromagnetic radiation. The acceleration limits can be far exceeded depending on the capture process.

The conversion electron that resulted from the capture of an orbital electron by the nuclear proton is also a source of fast electrons. However, all internal transformations are different and more complex than those deriving from the emission of beta particles (β). Basically, an electron lying at a low orbital zone being captured by the nuclear proton, which upon becoming a neutron, reduces the deuterotomic number of the element then formed. The electron situated at an upper orbital zone moves ahead to occupy the vacancy left by the electron captured. Example of the capture of an orbital electron by a nuclear proton:

Ce58/139 + e^- ———> La57/139 (conversion-electron energy = 126 keV)

The capture of an electron K by the nuclear proton occurs in the genetic pole of cerium deuterotome (Ce58) made up of noble gas xenon (Xe54). When the nuclear proton captures an electron K, it is changed into a neutron and reduces the deuterotomic number by one; next it is transformed into lanthanum (La57), keeping the mass number unaltered. The electron's displacement takes place from P shell to K shell and the loss of energy will give rise to spectrum lines, which is a pattern of the element.

A neutron being added to the deuterotome causes the unit's capacitance to rise, for the distance **d** of the dielectric increases. The capacitance alteration, however, is not limited to that unit; it affects the deuterotome's total capacitance by decreasing it. Because the deuterotome is a multiplate capacitor, the reduction of a plate shortens the overall area (A), decreasing its value.

Thus, the conversion electron–due to the capture of an orbital electron by the nuclear proton–causes the following changes:

1) The deuterotomic number (Z) is decreased,
2) The dielectric number is increased and, so is, its thickness (**d**),
3) The unit's capacitance then formed is decreased,
4) The total capacitance C of the deuterotome then formed is decreased,
5) It leads to one other type of electromagnetic radiation known as inner bremsstrahlung,
6) The binding energy per nucleon is decreased.

b) Beta decay. The internal conversion process mentioned previously is set in motion from an excited nuclear state that usually results from beta decay (emission of beta particles). Those radioactive elements that spontaneously emit beta particles (β) also produce an element of greater deuterotomic number in addition to emitting gamma rays due to the excited nuclear state.

The common mechanism that regresses the excited nuclear state of any element is based on the emission of one or more photons of gamma rays. In some nuclear state of excitations, the emission of gamma rays can be somewhat inhibited, that is, negligible, and the electronic excitation arising out of internal conversion can become significant. At this point, the energy of nuclear excitation (Eex) is transferred outright to one of the deuterotome's orbital electrons. This electron, then, comes up with an energy given by the equation: $Ee^- = Eex - Eb$, where Ee^- is the electron's energy, Eex the energy of nuclear excitation and Eb the binding energy in the original electronic shell.

The following are common sources of conversion electrons: 1) emission of beta particles (β); 2) electron capture by the nuclear proton. Below is an example of radioactive transformation through the emission of beta particles and the value of the conversion electron energy, expressed in keV.

Cs55/137 - β ——————>Ba56/137 (conversion electron energy = 624 keV)

Significant alterations also occur in the conversion electron process due to the emission of beta particles (β) such as:

1) The deuterotomic number (Z) is increased,
2) The thickness **d** in the dielectrics that make up the deuterotome is decreased,
3) The capacitance C of the unit then formed is increased,
4) The total capacitance C of the deuterotome then formed is increased,
5) It also gives rise to one other type of electromagnetic radiation, known as inner bremsstrahlung,
6) The binding energy per nucleon is increased.

11.2.4 – Auger electron

The energy of nuclear excitation produced by the capture of an electron can be transferred outright to an electron situated in an outlying electronic shell and cause it to be pulled out of the deuterotome. The electron thus ejected is called Auger electron. Another source of energy that can lead to the production of an Auger electron is the state of excitation of electrons in internal electronic shells. It can be transferred directly to an electron situated in an outlying electronic shell and cause its ejection; this electron is also an Auger electron.

This is a physical phenomenon per se, but the cause of excitation of electrons in internal electronic shells or, rather, the cause of excitation in the nucleus, precedes both the formation and the ejection of Auger electron. The initial cause can be either the electron capture by the nuclear proton–also a source of conversion electron–or the beta decay, also a process of nuclear excitation, both inverse processes; in other words, the capture of an electron and the emission of an electron, respectively.

The electron capture is a physical phenomenon that causes a significant alteration in the energy, either nuclear or electronic. Electronic excitation results from the vacancy left in the K shell by the capture of an electron K and the consequent need to reset electrons in the shells through the consecutive emission of characteristic X-rays photons;

conversely, nuclear excitation results from the capture of an electron. However, as the nucleus cannot maintain that state endlessly, it unleashes a process of deexcitation by emitting nuclear gamma rays.

Beta decay is also a physical phenomenon that causes a significant change in the nuclear energy as well as some electronic alteration. The electron emitted by the intranuclear neutron is ejected from the deuterotome, making up beta particles (β). Similarly to the electron capture, the excited nuclear state sets up deexcitation and emits nuclear gamma rays. During beta decay, the energy of nuclear excitation can also be transferred to an electron in the outlying electronic shell and cause its ejection (Auger electron).

The energy of Auger electron is given by the difference between the initial energy of deuterotomic excitation and the binding energy of electrons in the shell from where it was ejected. This electron has a discrete spectrum of energy with different elements and different initial and final states. In all these cases, its energy is relatively low compared to beta particles and conversion electrons, particularly because the emission of Auger electron is only favored in elements of low deuterotomic Z number, where the binding energy of electrons is low.

Beta decay can be symbolically represented as follows:

$$n \text{ -------> } p + e^- + \bar{\nu}$$

where **n** is the nuclear neutron; **p** the residual proton that made up the nuclear neutron; **e**$^-$ the electron emitted by the nuclear neutron; □**v** the antineutrino emitted due to the excited nuclear state. The nucleus deexcitation will cause the nuclear gamma rays to be emitted and the nuclear excitation energy to be transferred. Should there be total transference to the orbital electron, this will cause its ejection to be characterized as Auger electron. Two beta decay examples are shown below:

$$\text{Ce58/141} \xrightarrow{-\beta} \text{Pr59/141} + \gamma \text{ (+ Auger electron)}$$

$$\text{Po84/215} \xrightarrow{-\beta} \text{At85/215} + \gamma$$

A high deuterotomic number (Z) like that of polonium (Po84) does not work in favor of the emission of Auger electron. The illustrations sketched below show the microscopic aspect of electron capture and beta decay, with resetting electrons in the electronic shells.

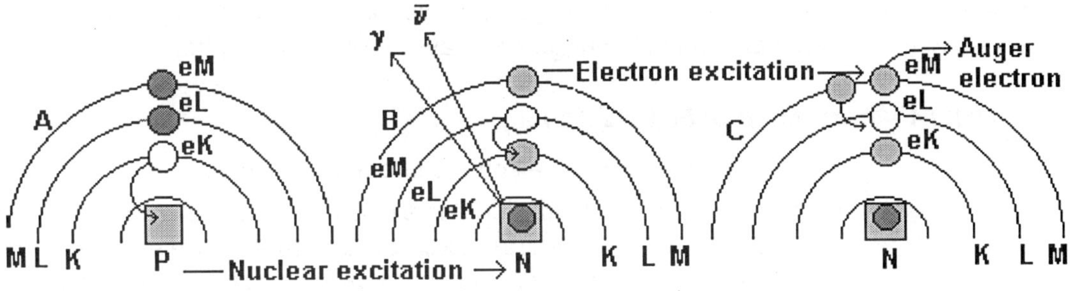

Fig. 90 - **Electron capture**. **A** – The nuclear proton captures electron K and leaves a vacancy in the shell, creating the initial electronic excited state. **B** – Electron L moves from K shell, with a lower energy level, emitting characteristic X-ray photon and occupies the vacancy left by the K shell; the nucleus gets excited as well and emits nuclear gamma rays and neutrino. **C** – The energy of the electron in excited state is transferred to electron M, causing its ejection (Auger electron).

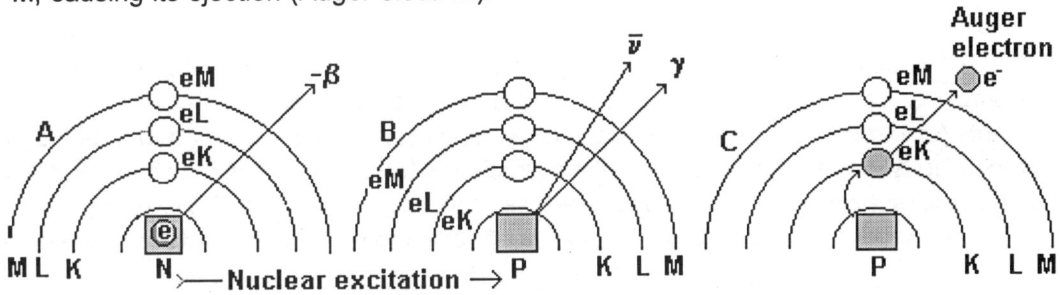

Fig. 91 - **Beta decay**. **A** – The nuclear neutron emits a beta particle (β), which is an electron with negative load. **B** – When the electron is emitted the neutron is converted into a proton causing the initial element to be transformed into another with a higher deuterotomic number; this physical phenomenon creates a nuclear excited state that results in the emission of gamma ray and antineutrino. **C** – The energy from the nuclear excited state can be transferred directly to an Auger electron, generally in the K shell that causes its ejection from the deuterotome.

11.3 – Radiation Interactions

The luminous effects that result from emissions of photons such as the fluorescence in the glass wall of Crookes' tube or the luminescence of radioactive cesium, for example, that appear in materials lying nearby or in emitting materials themselves, are entirely dependent on the physical phenomena that occur in the intramural or intradeuterotomic space reached by electrons and photons, and interact with deuterotomic structures depending on the route they may follow, either freely or colliding.

What is known about the interaction of radiations with matter plays an important role for the fabrication of instruments to detect radiations. Thus, the operation of any radiation detector depends on how the radiation to be detected interacts with matter of the detector itself. Two major types of particles must be taken into account in this field: those subject to Coulombian forces with electric load (alpha particles and fast electrons) and those not subject to the forces mentioned above (gamma rays, X-rays and neutrons).

11.3.1 – Interaction of fast electrons

The following are present when fast electrons interact with matter:

1) Loss of specific energy due to ionization and excitation. This loss may result from radioactive process and the radioactive loss takes on the shape of electromagnetic radiation, which can derive from any position along the irregular route taken by the electron. Radioactive losses always account for a small fraction of energy lost due to both ionization and excitation, significant only in absorbing materials with a great deuterotomic number (Z),

2) Even absorbing materials with small thickness may cause the loss of a few electrons through an absorption process of mono-energetic electrons coming from the bundle detected because the dispersion of electrons removes them from the flow that gets to the radiation detector,

3) The absorption of beta particles (β) also occurs in absorbing materials with small thickness because these particles are provided with low energy,

4) An electron entering the surface of an absorber can be submitted to a deflection process as soon as it emerges from the surface it entered. This physical phenomenon is called backscattering. The backscattered electrons do not deposit all the energy in the absorber medium, and warrants a significant effect in the response of radiation detectors.

11.3.2 – Interaction of gamma rays

A large number of mechanical processes is known in the interaction of gamma rays and X-rays with matter, but only three of them play an important role so that the radiation rate can be obtained, namely: photoelectric absorption, Compton dispersion and the production of a pair of electron-antielectron.

These three processes carry out the partial or total transference of energy from the gamma-ray photon to the electron's energy. This process of energy transference results in a fast and sudden change where the photon may have two destinies: disappear altogether or be dispersed through a sharp incidence angle.

a) Photoelectric absorption – A photon interacts with an absorbing deuterotome in this process where the photon disappears completely. In its place an energized photoelectron is ejected by the deuterotome of one of its orbital electron shells. The interaction occurs entirely in the deuterotome and does not happen to free electrons. As regards gamma rays of sufficient energy, the most likely photoelectron origin lies in the electronic shell situated more inwardly or in the deuterotome's K shell. The photoelectron's energy is given by the equation $Ee = h.v - Eb$ where **Ee** is the photoelectron's energy; **h** the constant of Planck (= 6.626×10^{-34} J.s.); **v** the radiation frequency and, **Eb,** the photoelectron's energy in its original shell.

This interaction also creates an absorbing deuterotome that was ionized by the loss of electrons in one of its shells through the capture of a free electron that comes from the surrounding environment or a realignment of electrons coming from other electronic shells of the deuterotome. Thus, one or more characteristics of X-rays photons can be likewise generalized. In many cases, these X-rays are absorbed again in the original site through photoelectric absorption. In some cases, however, the emission of an Auger electron can replace the features of X-rays, by taking the energy of electronic excitation out of the deuterotome. The Auger (French physicist, 1899) electron is the one emitted as a result of excitation of electrons in the electronic shell, the onset of which was caused by an electron captured by the nuclear proton, whose energy level of nuclear excitation was transferred directly to the electron in the electronic shell, the emission of nuclear gamma rays is inhibited.

The photoelectric absorption process is the predominating medium of interaction for gamma rays and X-rays of relatively low energy. The process is also significant in absorbing materials with high deuterotomic number Z (number of protons).

Potassium iodide (KI) is a material often used to detect gamma rays. In maps registering sections of photoelectric absorption by that substance, in areas with low energy, are gamma rays that correspond to energy levels connected to electrons in several electronic shells of the absorbing deuterotome.

As for energy of gamma rays slightly above the limit, the photon's energy is enough to withstand a photoelectric interaction where an electron K is ejected out of the deuterotome. For energy levels slightly below the limit, a photoelectric interaction is in no way energetically possible. Similar absorption limits occur in lower energy for the remaining electronic shells.

An electron belonging to the K shell of a certain element can, if stimulated, set off a state of excitation and then partially transfer its energy to another electron in the form of photons, thus making up a new unit called **photoelectron** with an energy level well above the original one. Of the three major physical phenomena that happen in the interaction of gamma rays, the most outstanding one is the absorption of a photoelectron by an absorbing deuterotome.

b) Compton scattering – This process takes place between a gamma-ray photon that strikes a target and an electron in an absorbing material. It is the most frequent interaction mechanism for energy of gamma rays in sources of radioisotopes.

In the scattering process the entrance of gamma rays photons is deflected through a theta (θ) angle in comparison with its original direction. The photon partially transfers its energy toward the electron, this transference is known as **recoil electron**. During scattering any deflection angle is possible. Thus, the energy transferred to the electron can vary from zero to a large fraction of gamma-ray energy.

c) Production of electron-positron – If the energy of gamma rays is twice that of the rest mass energy of an electron (1.02 MeV/c^2), the process of production of electron-positron pair (or electron-antielectron) is energetically possible. The pair production is mainly restricted to gamma rays with high energy. The rest mass is that of a body in absolute rest, when its speed is zero.

In the interaction likely to occur in the Coulombian field of a nucleus, the gamma ray photon disappears and is replaced with a pair of electron-positron. Accordingly, when a photon hits a deuterotome, it can lose its energy entirely and produce a pair of electrons. In fact, during this process of pair production two particles are created: an electron and a positron (positron is a particle whose mass is identical to that of the electron, with positive electric load though). The production of the pair of electrons results in energetic mass. The electron did not exist before the photon had hit the deuterotome; moreover, the electron was not a part of the deuterotome the photon had hit. The photon energy is transformed into total relativistic energy E^+ of the positron and E^- of the electron. The energy production can be expressed by the equation **$h.v = E^+ + E^-$**, where $E^+ = m_e c^2 + K^+$ and $E^- = m_e c^2 + K^-$, where

$$h.v = (m_e.c^2 + K^+) + (m_e.c^2 + K^-)$$

As the values of kinetic energy K+ and K– are always positive, the equation is now: $h.v = 2\ m_e.c^2 + K^+$, where **m_e** is the electron mass (equal to 0.5110 MeV/c^2), **c** is light speed (3×10^8 m/s) and **K^+** the kinetic energy of electrons.

Consequently, so that the process of electron pair production may happen it is necessary that the photon have energy of at least $2.m_e.c^2$ (1.02 MeV), that is, twice the energy of the electron. Photons with those energy levels are encountered in areas occupied by nuclear gamma rays. What happens can be symbolically represented as follows:

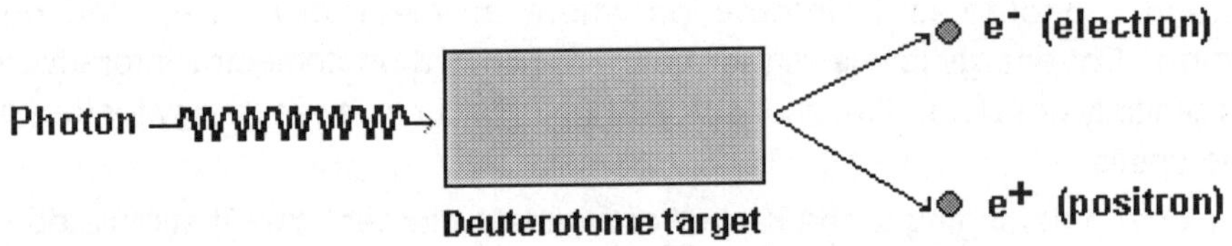

The opposite process, known as **positron annihilation,** can happen to free electrons and positrons. In this reaction, at least two photons are created. Both the electron and antielectron (positron) disappear and are replaced with two photons. The energy in these two photons are: E1 and E2. On the basis of equation $h.v = E^+ + E^-$, we have:

$$E1 + E2 = (m_e.c^2 + K^+) + (m_e.c^2 + K^-)$$

The momentum of conversion requires that the two photons have energy $me.c^2$ (equal to 0.5110 MeV) and that they move in opposite directions. The process can be symbolically represented as follows:

electron (e^-) + positron (e^+)————> 2 photons

Therefore, the importance of the three mechanisms referred to above lies in the fact that they govern the partial or total transference of energy in gamma rays photon to the electron's energy, taking into account the absorbing matter of photoelectrons.

CHAPTER 12

Atom: a new model

12.1 – Crookes' tube as deuterotomic model of radiation emitter

Experimental studies on the response of electric discharges in rarefied gases (around 1850 to 1897, approximately) were carried out inside a long-shaped glass tube. At one of the extremities was fixed an electrode called cathode and an electric load was applied to it; one other electrode called anode was fixed in the opposite extremity. The air inside the tube was blown out until the internal pressure reached 0.03 mmHg. The simplified scheme illustrates the major features of the tube and the position of the electrodes.

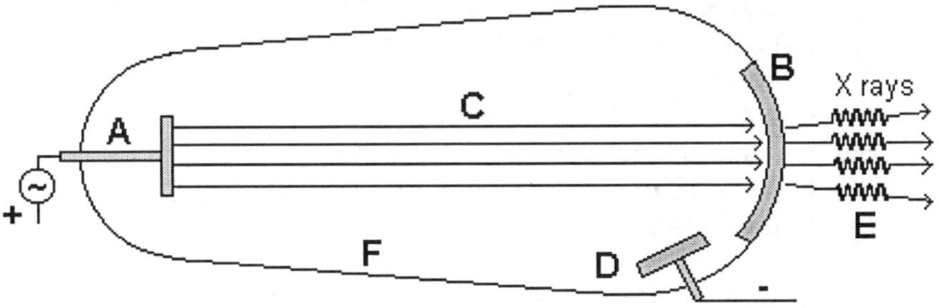

Fig. 92 - **Crookes' tube**: simplified scheme. **A** – Cathode of the tube where an electric load (from 10,000 to 30,000V) is applied. **B** – Target glass wall, where the cathode rays (accelerated electrons) will collide. **C** – Bundle of cathode rays. **D** – Anode of the tube, not necessarily the point of impact of electrons of cathode rays. **E** – Electromagnetic waves originating in the glass wall, called X-rays. **F** – Interior of the tube with rarefied gas and very low pressure around 0.03 mmHg.

By applying a high voltage electric discharge to the cathode, the discharge takes on a bluish brightness and a green fluorescence appears in the glass wall placed in front of and close to the anode. Because the electrode that receives the electrons load emits rays, they were named cathode rays by Goldstein in 1876. From the glass wall, which becomes fluorescent, leave rays to the exterior of the tube; not very well known at that time, they were called X-rays or Roentgen rays.

Cathode rays and X-rays differ in their properties. Bundles of cathode rays are swerved by the magnetic pole of a magnet, a physical phenomenon not found in X-rays. The cathode rays were identified as having been formed by accelerated electrons and X-rays as electromagnetic waves with high frequency and short wavelength. The penetration power of rays in wood, aluminum, meat, paper, etc. is huge. However, they are not capable of going through such metals as lead, platinum and calcium and bones or materials that contain calcifications. Nevertheless, X-rays have different penetration powers depending on the amount of residual air inside Crookes' tube; if the internal space contains a lot of gas, the continuous X-rays produced will be soft and with a little penetration capacity in matter; conversely, if the residual gas is scarce, the X-rays then produced will be hard with a great penetration capacity, therefore with a variable property.

Gamma rays photons have great penetration, similar to hard X-rays. Given the origin of hard X-rays–due to low concentration of gas in Crookes' tube–by analogy, it can be inferred that nuclear gamma radiations are formed inside a deuterotome, whose gaseous concentration is null. Since both X-rays and gamma rays are electromagnetic waves, a comparative study was carried out between Crookes' tube and a radioactive heavy deuterotome. The tube shape and its capability to produce radiation (X-rays) suggested the possibility of a heavy deuterotome (uranium, for example) to have a similar configuration, due at least to eight major reasons:

1) – It also emits radiation (gamma radiation),
2) – It requires a source of accelerated electrons,
3) – It requires an empty internal space to hurl its electrons,
4) – It requires an energetic pole emitting electrons,
5) – It requires a target wall, which the electrons can hit,
6) – It has to be more or less oblong shaped,
7) – It has to have a permanent energetic source of electrons,
8) – It fuels energy of nuclear excitation.

These comparative and theoretical characteristics are associated with the following:

1) The theory of an atomic capacitor where the atoms of deuterium and tritium would be those capacitors, units that make up deuterotomes from helium (He2) to uranium (U92),

2) The theory of a genetic pole in deuterotomes, formed by noble gases and the chemical pole, corresponding to the extremity where chemical reactions are processed,

3) The characteristics of series-equivalent bindings for deuterotomes up to xenon (Xe54) and parallel-equivalent bindings from cesium (Cs55) up to natural element radon (Rn86) made it possible to assume the existence of an "electronic furnace" inside a heavy deuterotome by using uranium as a template.

Thus, the analogy between Crookes' tube and a heavy deuterotome made it possible to set up a model consisting of five major parts: 1) genetic pole, 2) chemical pole, 3) "electronic furnace", 4) wrapper wall and 5) an empty internal space. Thus, this new atom model, now called **deuterotome model,** started being utilized to account for radioactivity, electricity, magnetism, chemical valences, the high density of noble gases, in addition to other physical properties. The expression **atom model** is exclusively intended for hydrogens (light hydrogen, deuterium and tritium).

Accordingly, the deuterotomic theory is based on the following data:

1) The quantum theory where a quantum of energy (photon) can be the same package of lentons themselves,

2) The emission of photons (quantum) when an accelerated electron strikes a target wall fits into the likely existence of a package of lentons in the electron's interstice to be ejected as a result of the collision, thus reducing the electron's kinetic energy that lost the said energy package. Thus, lenton and quantum resemble each other and the loss of an electron's kinetic energy is associated with the loss of quantum energy,

3) The atom of hydrogen model by Rutherford-Bohr and Thomson where the electron with kinetic energy revolves about the proton,

4) The concurrent electronic shells and the respective distances between the electron and proton, proportional to the square of the whole numbers from 1 to 7,

5) The concurrent zones of orbital electrons and their respective energy levels;

6) The concurrent bands theory for electrons of electric currents to be carried over (electrical conductivity).

Moreover, it provides new insights into the following knowledge:

1) The origin of chemical valences (positive and negative),

2) What the high density of noble gases stands for in the composition of deuterotomes,

3) The importance of one neutron participating in the structure of light hydrogen atom.

4) A new theory creating the figure of atomic capacitor and on the series-equivalent and parallel equivalent bindings between them,

5) A new theory creating the deuterotome with an internal space,

6) The capability of a genetic pole to adsorb neutrons on its surface,

7) A new theory creating the magnetic tunnel, mainly in iron, cobalt and nickel deuterotomes, the one of electron-electric and that of electron-magnetic as well as the one of solenoid electro-circulation in permanent magnets,

8) A new theory creating the "electronic furnace" formed by 32 electrons and a 32 eV electric load in uranium deuterotomes in an intradeuterotomic position in heavy radioactive elements,

9) A new theory creating the intradeuterotomic electron-projectile ejected from the "electronic furnace" against the deuterotome's chemical pole and the alpha appendix,

10) A new theory creating the alpha appendix of heavy radioactive deuterotomes, formed by an alpha-unit, the composition of which is $\alpha 2/4$, $\alpha 2/6$, $\alpha 2/8$, $\alpha 2/12$ whose "locus" determines the code for emissions of alpha ($\alpha 2/4$) and beta particles (β) in radioactive families.

Nevertheless, the assumption that the nucleus is a bunch of protons and neutrons in disarray is entirely ruled out (despite Mayer theory where these units are arranged in intranuclear proton-shells and neutron-shells). However, it acknowledges the valuable service that the electronic shell theory has rendered to the bands theory as regards electric conductivity; the production mechanisms of X-rays, gamma rays, Auger electron; the acknowledgment of a vacant space in the electronic shell for the electron capture by a nuclear proton. All of these are the basis for the entire awareness on the origin of nuclear and electronic excitation energies; the acknowledgment that electromagnetic waves are emitted due to the loss or gain of energy when the electron moves along energy levels; and the awareness of interactions of photons and electrons colliding with the target matter.

A simplified scheme of Crookes' tube (I) and a heavy radioactive deuterotome (II) is shown below by comparing the respective structures:

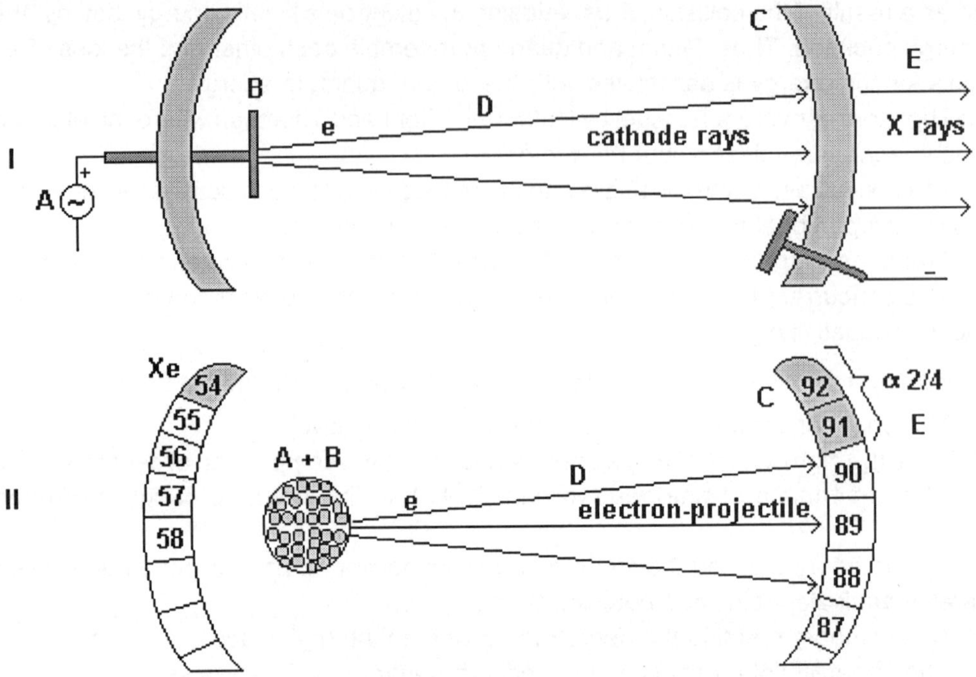

Fig. 93. **I) – Crookes' tube**. **A** – Source that generates the electric load to be supplied to the cathode (10,000 to 30,000 V). **B** – Tube's cathode, which is the source that emits cathode rays. **C** – Tube's target wall, made of glass, where the cathode rays will collide. **D** – Bundle of accelerated electrons emitted by the cathode formed by electron-projectiles. **E** – Electromagnetic waves (photons of continuous X-rays) are taken to the outside of the tube resulting from the collision of the electron-projectiles with the target wall. **II) – Heavy radioactive deuterotome. A + B** – "Electronic furnace" formed by 32 electrons with a 32 eV potential, at the same time the source that generates energy and the pole that emits electrons, close to the genetic pole composed of noble gas xenon (Xe54). **C** – Deuterotome's target wall formed by the alpha appendix, at the same time its chemical pole. **D** – Electron emitted by the "electronic furnace" and making up the electron-projectile, which will be thrown against the target wall (that will cause the rupture of the proton/neutron junction and consequent emission of alpha particles). **E** – Emission of alpha particles (α2/4) led to the outwards of the deuterotome.

The following can be seen in Crookes' tube scheme:

1) The source of electrons (cathode) where the electric load is applied,

2) The glass screen where the electron-projectiles emitted by the cathode hit each other,

3) Glass wrapping that shapes up the tube and makes it possible for the gas to be removed from the tube, thus reducing the internal pressure,

4) Tube's internal space where the molecules of gases are rarefied and can be ionized due to the action of electrons.

The following can be observed in the heavy deuterotome's (U92) scheme:

1) The source of electron-projectiles that consists of 32 electrons in the P shell with a 32 eV potential, called "electronic furnace",

2) Screen formed by the structure of deuterons in the chemical pole and the deuterotome's alpha appendix, where the electrons emitted by the "furnace" collide,

3) Deuterotome's wrapping formed by the sequence of deuterons that are round- and spiral-shaped and maintains the wall structure, preventing the reduction or modification of its internal space,

4) Internal space where the original condition of "empty space" is maintained.

The wall of deuterons that make up both the chemical pole and alpha appendix is the shock-absorbing structure of electron-projectiles that originates from the "electronic furnace" of heavy radioactive deuterotomes. Even though this wall is subject to shocks, its thickness is small. When the electron-projectile strikes the proton-neutron (P/N) junction of the alpha appendix, it causes its rupture and triggers the emission of alpha particles ($\alpha 2/4$); the resulting energy from nuclear excitation that is installed can lead to emissions of nuclear gamma rays as a process of nuclear deexcitation. Should the energy thus created, however, fail to be adequate, the emission of gamma rays then does not take place; if the energy from the nuclear excited state should be transferred directly to an electron in the outlying electronic shell, emissions of Auger electrons will occur and the emissions of nuclear gamma rays will be impaired.

The energy from nuclear excitation will be higher whenever it arises out of emissions of beta particles (β) and the emission of nuclear gamma rays will almost be a constant, accompanying beta emission; beta decay is not contingent on the existence of the "electronic furnace" because elements of Z deuterotomic number below 53 can emit beta particles (β), for they are devoid of an "electronic furnace." Alpha decay depends on the existence of the "furnace" because only heavy deuterotomes are capable of emitting alpha particles ($\alpha 2/4$) spontaneously.

A simplified scheme of a heavy deuterotome with its major elements shows the emission mechanism of alpha ($\alpha 2/4$) and beta (β) particles.

The electric load applied to the cathode of Crookes' tube is inside of an open system; in other words, there is a source for emission of electrons (cathode) and a pole for receiving of these electrons (anode). Although the electrons emitted by the cathode do not go straight ahead towards the anode, they will be captured after the collision with the glass wall close to the anode and ejected out the tube. The electrons do not build up in the glass wall they strike; an electron traveling the route from the cathode to the glass wall situated in the opposite extremity hits the glass wall, loses kinetic energy and emits a continuous X-ray photon.

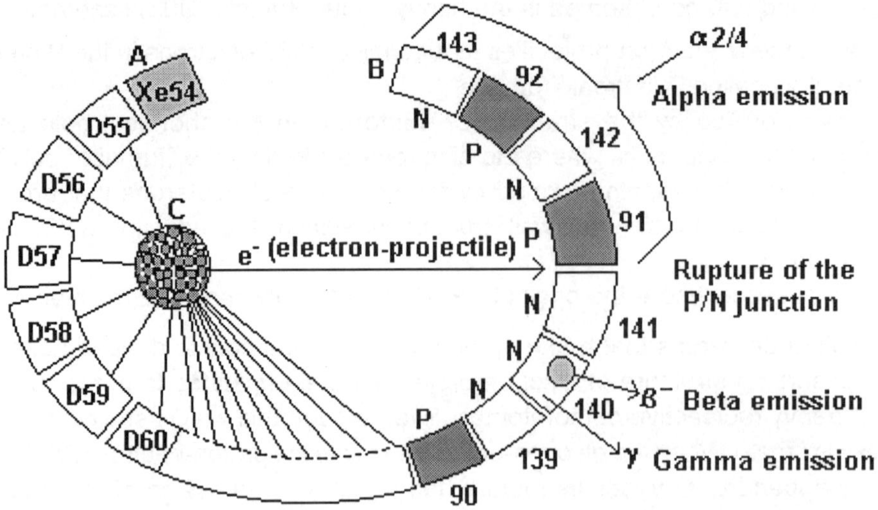

Fig. 94 – **Simplified scheme of a radioactive deuterotome (U92/235)**: 1) The "electronic furnace" C emits an electron-projectile that reaches the proton/neutron junction (P/N) in the free extremity ("alpha-appendix"). 2) The rupture of the P/N junction separates the unit formed by two protons and two neutrons that corresponds to the alpha particle (α2/4). 3) Following alpha decay, neutron n-140 is ready to emit a beta particle (β); beta decay causes an initial excited nuclear state; deexcitation goes on the final state, and is accompanied by the emission of nuclear gamma rays. 4) In the genetic pole formed by xenon (Xe54), the electron-K captured by the nuclear proton of that structure will also produce an excited state in nuclear and electronic levels alike; the nuclear deexcitation will produce emissions of nuclear gamma rays and the rearrangement of electrons to occupy the vacancy left as a reason of electron-K capture, moving from shells of higher energy levels to lower levels, will cause the emission of characteristic X-rays photons of the element and, likewise, the emission of Auger electron, depending on the energy of the excitation state.

The electrons that split the proton-neutron junctions in the alpha-appendix of heavy deuterotomes (uranium-238, for example) have an internal origin, that is, they belong to the "electronic furnace", having therefore the power to break apart the P/N junctions. Although the electrons in the cathode rays of Crookes' tube also hit the P/N junctions, they have, however, an external origin (extranuclear) in relation to the deuterotome. For that reason, they are unable to produce the same type of fission as occurs on radioactive elements.

The electric load existing inside the uranium deuterotome (U92) belongs to a restricted system. The "electronic furnace" is the natural and permanent source for emission of electron-projectiles, without the corresponding anode that would work as a receptor of electrons. The system would fail to be viable should the intranuclear electrons be led outwards of the deuterotome, for the electric load would be depleted. The electrons emitted by the "furnace" strike the deuterium wall lying in the chemical pole and return to the "furnace", thus maintaining the 32 eV electric load, even if passing first by thorium (Th90), then radium (Ra88) and finally radon (Rn86); only from radon onwards does the electric load in the “electronic furnace” begin to fall, starting from 30 eV in polonium (Po84) to reach 28 eV in lead (Pb82) where the radioactive emission ceases for isotopes with mass numbers 206, 207 and 208. The electrons that strike the deuterons wall diagonally are deflected and tend to hit several other targets before returning to the “furnace”, depending of the incidence angle at the point of impact.

In the degradation mechanism of radioactive deuterotomes, the emission of beta particles (β) and nuclear gamma rays happens if they were concomitant. Alpha (α2/4) and beta (β) particles cannot be emitted simultaneously by the same deuterotome and the emission of gamma rays is a physical phenomenon that ensues after beta emission has occurred because the energy of nuclear excitation is a consequence of a break in the electric stability between the proton and electron that make up the neutron; the nuclear deexcitation gives off energy for gamma radiation in the form of electromagnetic waves, formed by the nuclear gamma ray photon.

During the synthesis mechanism known as “neutron capture, emission of beta particles” that happens when uranium-238 (U92/238) is bombarded by fast neutrons, the beta decay is present, though not dependent on the electrons in the “furnace.”

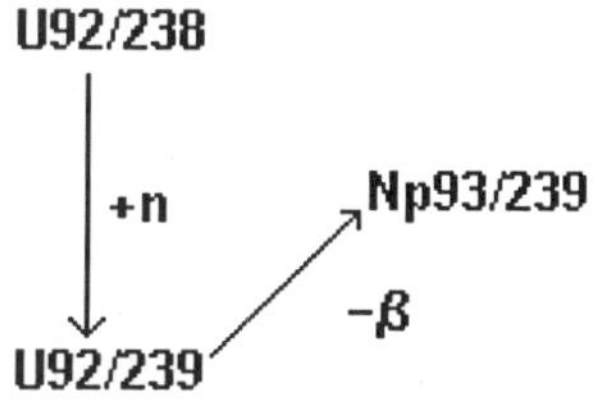

The emission of beta particles (β) can be contingent on the electrical instability produced by a neutron entering (dielectric) the center of two other neutrons, thus altering both the unitary capacitance and the total deuterotome’s capacitance.

Beta emission sets up an interruption in the synthesis progression by this mechanism in a great deal of deuterotome lineages. A few elements resulting from a post-beta decay are still receptive to capture a new neutron and generate one other isotope with greater mass number. This type of mechanism can be seen in deuterotomes with smaller Z deuterotomic numbers as shown in the following reactive sequence (* = radioactive):

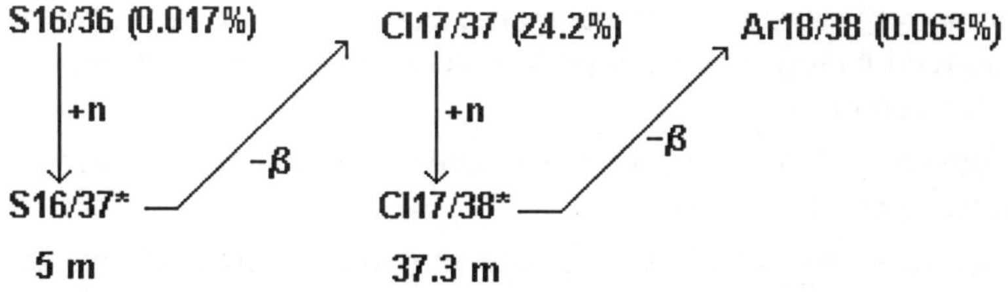

which is suggested as a likely way for the deuterotome synthesis (see Picture) where sulfur-36 (S16/36) fits into the characteristics of an “extinct isotope” or “residual isotope” because its concentration in nature is minimal (0.017%). There is no “electronic furnace” in sulfur deuterotome. Consequently, the beta emission is in no way dependent on any action by the electron-projectile of the "furnace." Thus, beta emission happens to elements with small Z deuterotomic number (Z16, for example), as well as to those with a great number (Z88, for example). However, alpha emissions only happen spontaneously in heavy deuterotomes.

12.2 – Uranium deuterotome: model of natural radioactive element

Gamma radiation is an outcome of radioactive property, in addition to others. Some of its physical properties are similar to those of hard X-rays. Both nuclear gamma rays and hard X-rays are electromagnetic waves devoid of electric load and mass, with no electric or magnetic nature whatsoever, nor electric fields or magnetic ones attracting gamma rays and X-rays. Both are capable of getting across intermatter spaces for they cannot be easily intercepted. Nevertheless, a few elements such as barium, calcium, lead etc, can block X-rays, whether hard or soft X-rays; this is more often observed in the latter, though. Gamma ray can cross the human body, as can X-ray.

Thus, a fundamental morphologic and functional likeness exists between Crookes' tube and a heavy deuterotome, uranium, for instance. The analogy between both of them made it possible to analyze in depth and simply, a number of physical, chemical, electric, magnetic and radioactive phenomena. When the atom model by Rutherford-Bohr and Thomson is employed for the whole range of elements, a few points still remain in the dark all related to phenomena originating from inside the nucleus. As far as light hydrogen (H1/1) is concerned, all goes well, but the same thing does not happen to deuterium, tritium and natural deuterotomes (from helium to uranium), by taking into account the proposed theory on atomic capacitors (with one or two dielectrics). In the model by Rutherford-Bohr, the nucleus set up is an extremely dark spot, failing to supply data for satisfactory comprehension because the current atomic theory does not suggest any organization regarding alpha emission apart from the Theory of Mayer that hints at a nucleus with a proton and neutron structure in concentric shells. The physical phenomenon of radioactivity itself, which acknowledges that alpha particles leave the interior of the nucleus, finds no microscopic explanation, being simply based on phenomena. The internal nucleus set up, in light of the classic atomic theory is unthinkable. The emission of beta particles themselves by a nuclear neutron, especially during the degradation of radioactive families, fails to provide a clear enough supportive explanation. However, the emission of nuclear gamma rays that result from nuclear excitation energy, due either to beta decay or the electron-K capture by a nuclear proton has nothing to do as to whether the atomic nucleus is organized or not.

The proposed deuterotomic model is based on a number of fundamental points of which the following can be mentioned:

1) The neutron is a dielectric or insulation that makes up the atomic capacitor, lying between two plates, where one is the proton and the other the electron.

2) The structure formed by the proton/neutron/electron, in that order, creates the basic capacitive atomic unit or unitary atomic capacitor.

3) Pairs of atomic capacitors (deuterons) are series-equivalent connected from helium deuterotome (He2/4) to those of xenon (Xe54).

4) The atomic capacitors (deuterons) are parallel equivalent connected from cesium (Cs55) to radon (Rn86).

5) Deuterotomes that reach the status of a noble gas, if provided with series-equivalent bindings in their own shell, go through the physical volume-reduction phenomenon, which confers high density upon each noble gas then formed. Should the bindings be parallel equivalent there will be no physical volume-reduction phenomenon; the capacitance becomes high with this sort of binding, thus preventing the volume-reduction phenomenon from taking place; an agglomerate of 32 electrons close to the deuterotome's genetic pole is then formed, making up the "electronic furnace" with a 32 eV potential.

6) The synthesis of deuterotomes from francium (Fr87) to uranium U92) is carried on by capturing alpha particles and neutrons and emitting beta particles, but xenon (Xe54) remains as their genetic pole. There is no theoretical evidence that radon (Rn86) is the genetic pole for francium, radium (Ra88), actinium (Ac89), thorium (Th90), protactinium (Pa91) and uranium (U92); otherwise there would be no "electronic furnace."

7) Noble gases, as deuterotomic capacitors, form the genetic pole for deuterotomes in each series, from K to O shell series.

8) During the synthesis process each deuterotome forms its own chemical pole, save for noble gases that do not have them. In this pole a number of deuterons will be available with free valences and an electromotive force (emf) determined by the series so that chemical reactions can happen. The potential difference between the valences of different deuterotomes and their chemical fitness determine the chemical reaction capacity.

9) The "electronic furnace" is an intradeuterotomic unit that triggers the physical phenomenon of radioactivity, specifically in the emission of alpha particles ($\alpha 2/4$).

10) Both the form and function in this new deuterotomic model are similar to those of Crookes' tube.

11) The repetition of synthesis mechanisms for deuterotomes falls into the types "alpha particle capture", "neutron capture" and "neutron capture and emission of beta particles ".

12) The finding of both "extinct isotopes" and "residual isotopes", the latter existing minimally in nature, as shown in the "Table of possible ways for the synthesis of deuterotomes".

13) The theoretical explanation as to the progressive incorporation of neutrons to deuterotomes during the synthesis of matter starting from thallium (Tl81/202, Tl81/203 Tl81/204), causing them to be heavy and radioactive and reach a mass number nearly 2.5 times higher than that of protons (Z deuterotomic number).

14) The theoretical explanation as to the existence of a great number of isotopes in a few deuterotomes such as samarium (Sm62) and xenon (Xe54).

15) The alpha-appendix identified in heavy deuterotomes with a phylogenetic code capable of maintaining constant those radioactive features in the families of radium, actinium, uranium-235 and thorium, as well as characterizing the composition of alpha-units and their "loci."

16) The theoretical evidence of a phylogenetic lineage in deuterotomes during the processes of synthesis.

17) The linear arrangement in atomic capacitors (deuterons) making up a round-shaped and elliptical-shaped deuterotome, somewhat like a spindle, thus making it possible to identify the following:

a) The existence of a genetic pole and a chemical pole in opposite extremities;

b) The intradeuterotomic position of electrons;

c) The series-equivalent bindings in deuterons from helium (He2/4) to xenon (Xe54) and from cesium (Cs55) to uranium (U92) in parallel equivalent ones;

d) That the 32 electrons available in the P shell make up the "electronic furnace."

12.3 – Genetic pole and Chemical pole of deuterotomes

Deuterotomes with longer structures such as bromine (Br35), iodine (I53) and uranium (U92) are provided with a more typical ring and elliptical shape compared to deuterotomes with shorter structures such as aluminum (Al13), sodium (Na11) and potassium (K19). All of them, however, have well-defined genetic pole and chemical pole. The genetic pole is formed by a noble gas that has gone through the physical volume-reduction phenomenon, whose volume is similar to that of the atom of hydrogen, save for neon (Ne10) and argon (Ar18), the volume of which is twice as much. The genetic pole is a source that generates tension. Each genetic pole leads the synthesis of matter at the time it is formed and establishes, partially, the physical and chemical properties of deuterotomes with the aid of its electric load. The electric load in the genetic poles is not similar to all them; instead, it depends on the number of receptor plates (protons) available within. Thus, with diversified number of plates, the capacitance values are also different.

The chemical pole is formed by the last deuterons following the noble gas that makes up the genetic pole. The chemical pole is the deuterotome area where the two major manifestations occur, namely: 1) the chemical reactions resulting from free valences and the ionization of polar molecules if they are put in a liquid or gaseous medium, 2) the radioactive degradation of heavy deuterotomes in the families of radium, actinium, urânium-235 and thorium.

12.4 – "Electronic furnace" of radioactive deuterotomes and radioactivity

12.4.1 – The origin of uranium "electronic furnace"

The physical volume-reduction phenomenon occurs in noble gases whose pairs of deuterons are reciprocally connected in series-equivalent bindings, where the number of deuterons in that series is complete. This happens to deuterons in the K, L, M, N and O shells where the noble gases then formed are He2, Ne10, Ar18, Kr36 and Xe54, respectively.

From xenon onwards, the synthesis of matter by way of the "alpha particles (α2/4) capture" in the phylogenetic lineage of U92, the deuterons that make up the structure ranging from D55 to D86 are not series-equivalent connected but, rather, parallel equivalent. Accordingly, those deuterons that make up the series in the P shell from xenon (Xe54) to radon (Rn86) in uranium deuterotome are parallel equivalent connected.

The 32 protons are arranged as if they were one single plate (receptor) and the respective electrons, the other plate (negative). A simplified scheme shows the bindings between protons and between electrons.

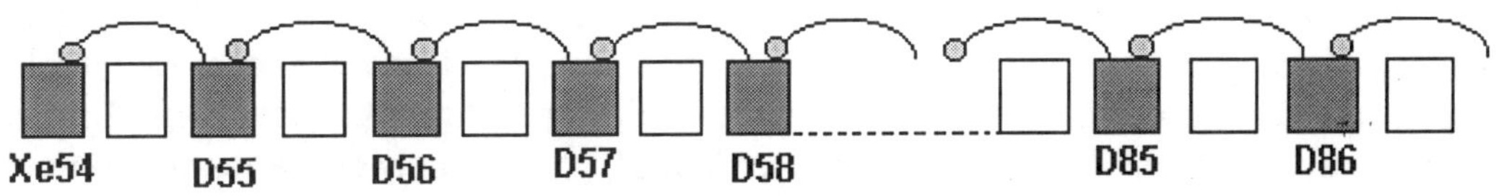

Fig. 95 – Simplified scheme of parallel equivalent bindings where the electron of a proton is linked to the previous unit. The electron of proton D55 is linked to the genetic pole formed by xenon (Xe54).

The "electronic furnace" was not formed until francium (Fr87) was made up. When that element was finally available (with a deuterium and an electron of valence, post-noble gas) and, further, the ensuing synthesis in elements with greater Z deuterotomic number, the "electronic furnace" was then formed, the 32 electrons being displaced to a point inside the deuterotome close to the genetic pole; the area where the electrons agglomerated, thus forming the "furnace", is the result of the proton-electron distance in the P shell. A simplified sketch of this structure can be shown as follows:

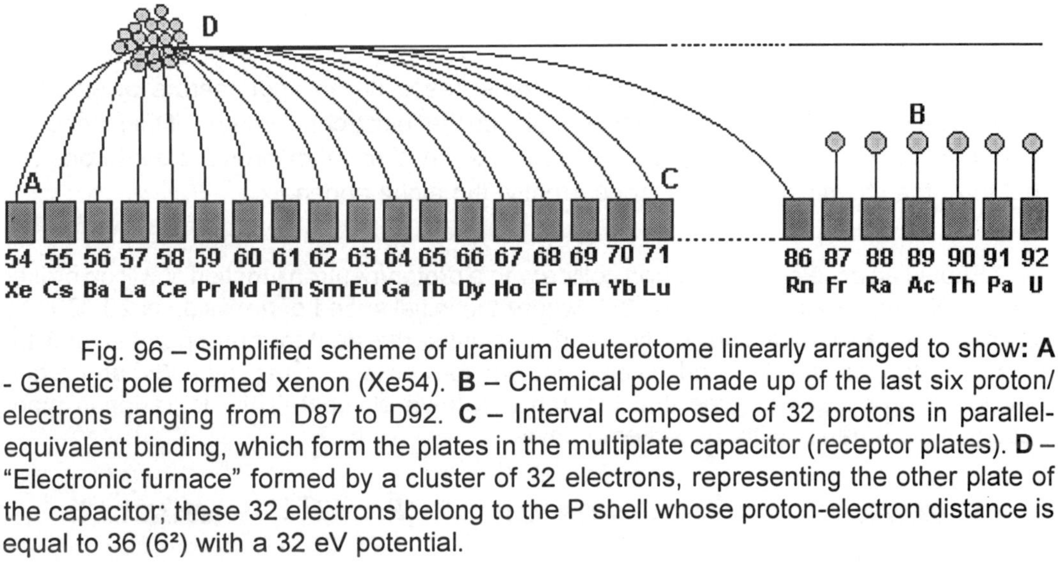

Fig. 96 – Simplified scheme of uranium deuterotome linearly arranged to show: **A** - Genetic pole formed xenon (Xe54). **B** – Chemical pole made up of the last six proton/ electrons ranging from D87 to D92. **C** – Interval composed of 32 protons in parallel-equivalent binding, which form the plates in the multiplate capacitor (receptor plates). **D** – "Electronic furnace" formed by a cluster of 32 electrons, representing the other plate of the capacitor; these 32 electrons belong to the P shell whose proton-electron distance is equal to 36 (6^2) with a 32 eV potential.

The structure of the 32 deuterons in the P-shell series is very long and the bindings among them are parallel equivalent, which prevents the structure from initiating volumetric reduction when the synthesis reaches the position occupied by radon (Rn86). The capacitance in that interval is high, once in parallel bindings the capacitances are added up. In parallel-connected capacitors, the total load (Qt) is the algebraic sum of individual loads (Q1, Q2, Q3, Qn, etc.). Thus, the equation Qt = Q1 + Q2 +.... Q32 can be replaced, since it is known that **Qt = Ct x Et** where **Ct** is the total capacitance in the P-series, **Et** is the electric load or total tension of the capacitor in the P-series expressed in volts. **Qt** is the total electric load expressed in Coulomb (C). **Ct** is expressed in Farad.

The parallel equivalent binding confers a great value upon the total capacitance of uranium deuterotome, and then, maintain the "electronic furnace" which is an inductor of radioactivity for emissions of alpha particles ($\alpha 2/4$); it does not participate in the emission of beta particles (β), though. Should the pairs of deuterons from D55 to D86 be series-equivalent connected, as is the case of the series K, L, M, N and O-shells, forming the "electronic furnace" would not be possible and, consequently, there would be no radioactivity by emissions of alpha particles. Matter would simply be dependent on beta and gamma radiations, in addition to emissions of discrete X-rays, characteristic of elements, as natural radioactive processes and the composition (percent) of elements available in nature would be quite another.

The participation of neutrons in the synthesis of deuterotomes by the "neutron capture" mechanism in association with the "capture of alpha particles" cause heavy deuterotomes to be formed; deuterons from D55 to D86 are incapable of being series-equivalent connected. Thus, the parallel-equivalent bindings become imperative due to the presence of two or more neutrons between two protons.

12.4.2 – The origin of alpha and beta particles and nuclear gamma rays in radioactive deuterotomes

Because uranium (U92/238) deuterotome keeps up an agglomerate of electrons within (32 electrons), which, put together, make up the "electronic furnace", it then becomes a permanent source of accelerated electrons that are thrown towards the deuterons wall that form the chemical pole and, concurrently, the alpha appendix.

Should an accelerated electron collide with a proton/neutron junction, it will be broken apart releasing an alpha particle (α2/4) where the initial speed of emission is 20,000 km/s. That rupture will diminish, by two and four units, the deuterotomic number and the mass number, respectively, and the element then formed is thorium (Th90/234) and setting free the two or four valences that thorium can have at most. This degradation stage occurs in accordance with the scheme below:

$$\text{U92/238} \xrightarrow{-\alpha\ 2/4} \text{Th90/234} \xrightarrow{-\beta \quad -\gamma} \text{Pa91/234}$$

In the physical phenomenon of radioactivity, alpha decay alone depends on the "electronic furnace". Beta decay is a radioactive phenomenon that happens under several conditions; in the radioactive degradation of U92/238, U92/235 and Th90/232 and the process of matter synthesis by the mechanism "neutron capture, emission of beta particles". The radioactive phenomenon resulting from this beta emission is present in the Z deuterotomic number smaller than 54, an evidence that it is in no way dependent on the "furnace", as in the example below:

$$\text{Cd48/112} \xrightarrow{+n} \text{Cd48/113} \xrightarrow{-\beta \quad -\gamma} \text{In49/113}$$

Among many possible causes determining beta emissions, the capacitance value can be one of them. A capacitance value below the necessary can trigger beta decay in any of the conditions mentioned previously. The excess of neutrons forcing a captured neutron to be placed in a position close to another neutron or between two neutrons existing in the deuterotome structure decreases the capacitance value. The scheme below represents this assertion.

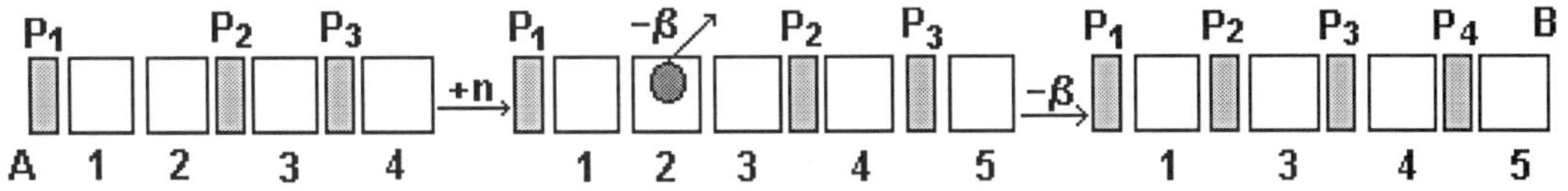

Fig. 97 – Schematic representation of “neutron capture and emission of beta particles”. where P represents the protons and the numbered pictures the neutrons. When a deuterotome captures a neutron, as shown in scheme A, it is placed between neutrons 1 and 2 and remains with five neutrons (scheme in the middle). Neutron 2 in the scheme in the middle emits a beta particle and is transformed into a proton represented by proton P2, placed between neutrons 1 and 3 in scheme B.

12.5 – Multiple dielectrics: one of the possible causes of beta decay

A neutron will not emit a beta particle (β) without a stimulus. In the neutron capture two energetic alterations are identifiable: a) reduction in the capacitance value; and b) reduction in the binding energy per nucleon. When a neutron enters the deuterotome, a sequence of 3 to 6 neutrons can ensue. This fact reduces the unit's capacitance and enables the neutron to emit an electron. If the neutron emitter fails to be ideally positioned between two other neutrons and if the capacitance is not below a satisfactory value, then the requirements for beta emission will not be met.

The phylogenetic code available in the alpha appendix of deuterotomes that make up the radioactive families of radium, actinium, uranium-235 and thorium complies with those two requirements. In the radioactive families mentioned above, deuterotomes can be found that give rise to alpha and beta particles, although the said emissions cannot take place simultaneously. This fact gives one to understand that within an amount of each element there must be a fraction that will first emit alpha particles and, then, beta particles (called alpha-fraction) and the other fraction will do just the opposite, that is, it will first emit beta particles and, then, alpha particles (called beta-fraction).

These fractions are present in Bi83/214 (family of radium), in Bi83/210 (family of actinium), in Bi83/211 and Ac89/227 (family of uranium-235) and Bi83/212 and Po84/212 (family of thorium). The two emissions (alpha and beta) cannot be carried out by the same deuterotome, for by the time one of them happens the element then formed will no longer be the same. Thus, the existence of these two fractions justifies emissions by the same element. The scheme below illustrates the mechanism of these two emissions. Figure 69 provides further comments to the above.

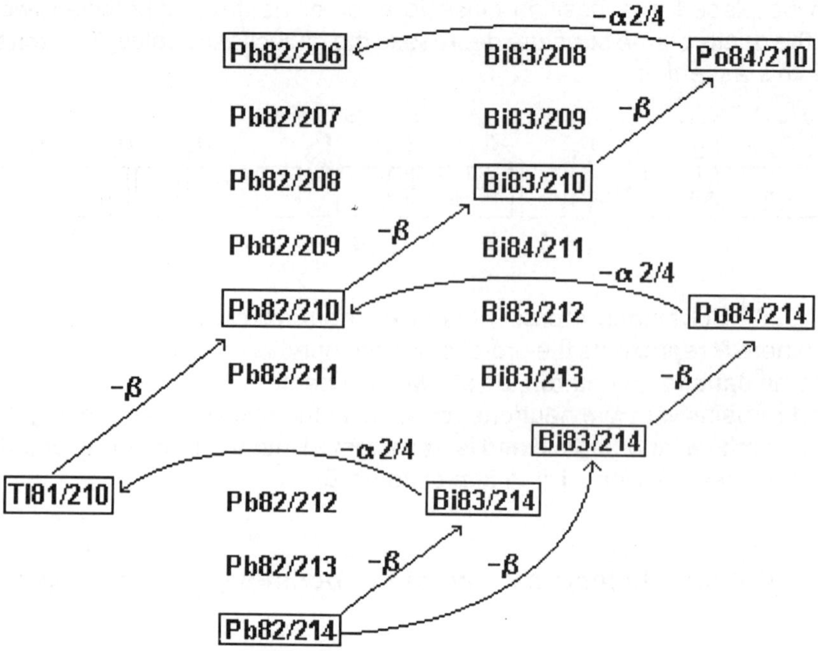

Fig. 98 – **Beta and alpha decays of Bi83/214**. In a certain amount of Bi83/214 there is a fraction that will emit a beta particle that takes precedence. After that it will emit an alpha particle (called beta-fraction of bismuth-214); the other fraction will do the opposite (called alpha-fraction of bismuth-214). The alpha emission will conduct that element on to Tl81/210 while the beta emission onto Po84/214; both will converge to Pb82/210 though, which will move on to stable Pb82/206.

The mechanism is identical in all elements provided with alpha and beta fractions within the radioactive families that can be seen in figures 69 to 72 and figures 61 to 64. The four figures (69 to 72) show the "loci" of multiple dielectrics (neutrons) and their individual beta emissions.

12.6 – Thermoradioactives properties of radium and its physical, chemical and biological effects

Radium (Ra88) is a radioactive element provided with properties and effects that stand out from most of the others, mainly in that it can release heat on a continuum, always giving the surrounding air a temperature surplus of 1.5°C. Radium as an element is more rarely found than uranium, for 3,270 kg of pure uranium are needed to obtain 1 gram of pure radium and it is more radioactive, approximately a million times higher than uranium (U92).

The emissions of alpha and beta particles and gamma rays photons by radium, starting from the ore that contains it (pitchblende), occur due to the presence of two isotopes, Ra88/226 and Ra88/228, the lighter emitting alpha particles and the heavier, beta particles and gamma rays.

During the process of radioactive degradation of elements in the radium family, Ra88/226 emits alpha particles (α2/4), then moving on to radon (Rn86/222). In the process of radioactive degradation in the thorium family, Ra88/228 emits beta particles (β), moving on to actinium (Ac89/228). Thus, each of these radium isotopes has a specific emission. Radium-226 or 228 alone cannot emit these two types of particles whether simultaneously or not.

The half-life of radium-226 is 2,205 years, not as long as that of uranium-238, which is 4,5 x 10^9 years and not as short as that of polonium-218, i.e. 4.3 minutes. That means that radium-226 is approximately 2 million times faster than uranium-238 during the process of radioactive degradation. Radium-226 is an alpha emitter, which hints at the fact that the electronical activity is very intense in the "furnace", and consequently very great the number of electrons that collide with the proton/neutron junctions. In addition to this great number of electrons that strike the P/N junctions, are the collisions resulting from deflexions by oblique incidences that are to occur continuously. Thorium-232 has a longer half-life than uranium-238, which is 2.2 x 10^{10} years. Other deuterotomes have shorter half-life, of days, hours or fractions of a second.

Radium salts ($RaBr_2$) have diversified effects, of physical, chemical and biological nature, of which the following can be mentioned:

1) Luminous effects: they are phosphorescent and provoke the same property in a great deal of substances set close to them, chiefly in glass, quartz, diamond, barium platinocyanide, and alkaline-earthy sulfides (Na_2S, K_2S etc.). Glass takes on a brownish or violet color; transparent quartz blackens, natural sapphire becomes yellow (a useful feature to distinguish a precious stone from a false one).

2) Electric effects: they ionize the air and other gases as well as those liquid that are poor conductor of electricity such as benzene, carbon sulfide (CS_2) etc., or even oxygen and nitrogen dissolved in them: N_2 <======== > $N^{\cdot} + N^{\cdot}$.

3) Calorific effects: a) they release heat continuously at 1.5°C above the environment temperature or the surrounding air, neither falling nor rising, on a continuum. Their value has been unchanged ever since that property was first identified until today; b) the radiations released by radium and absorbed by lead and the whole energy released is transferred in the form of heat which can be transformed into work; a gram of radium releases 135 calories-gram per hour, while coal, just for the sake of comparison, releases 8,000 calories-gram per hour.

4) Chemical effects: the radiations released by radium salts can produce a wide range of chemical effects: white phosphorous is transformed into its red allotropic variety; oxygen is transformed into ozone; nitric acid (H_2NO_3) is decomposed as if it were under the action of solar light; a radium salt, dissolved in water, decomposes the water molecule, slowly, in its composites: H_2 and O_2 at the rate of 10 cm^3 a day per every gram of radium.

5) Deleterious and pathological effects on animal tissues and vegetables, as well as unicellular beings: it destroys the germinating power of seeds; causes the leaves of vegetables to turn yellowish due to the degradation of chlorophyll; it causes the death of bacteria; it decreases or modifies the somatic development of animals growing up; it destroys the cells that present intense mitotic activity such as the embryonic and cancerous cells

CHAPTER 13

Electricity and Magnetism

13.1 – Electricity

The electric current is a flow of electrons that trickle through a conductor wire. In superconductors made of metal alloys, in pure metals or ceramics, the electrons roam about freely, there being no resistance to their passage nor a conductor system. In conductor and semiconductor metals, the electrons making up the flow are conducted through a carriage system known as **valence band** and **conduction band**, where the interval between them, known as **band gap** or, simply gap is provided with an amount of energy that will determine whether the metal is a conductor or an insulator. If the energy gap (Eg) is equal to or less than 1.1 eV, it features a metal conductor; conversely, if the energy gap is equal to or over 1.45 eV, it features an insulator for electric current or electrons.

The source of electrons can be either a potential difference in a system of plates (as in a battery, for example) or a transformation of magnetic electrons into electro-electrons in dynamos or magnetos when the armatures rotate inside the magnetic field, nearly juxtaposed to the N and S poles, and the electrons that leave the magnetic tunnel are captured by collectors and then led through the conductor wire as far as the delivery point (capacitors, resistors, etc).

If the tension in the source of electrons is too high, its distribution can be controlled by means of transformers. Conductors and semiconductors be resistant to the passage of electric current, which is different for each type of metal: for silver, 1.629; for copper, 1.724; for gold, 2.44; for aluminum, 2.828 at 20°C, the unit being expressed in $\mu\Omega$ (micro ohm); metals with small electromotive force (emf) also provides little resistance to the passage of electric current.

An armature (spire) that rotates between the N and S poles of a magnet and nearly juxtaposed to those poles will capture those electrons that escape from the magnetic tunnel of the magnet. The armature goes through a positive and a negative reversal, forming a senoidal wave. A complete wave corresponds to a cycle. The frequency of one cycle per second corresponds to one Hertz. If the system is formed simply by a magnet and, therefore, with an N pole and an S pole, in order for the electric current to have 60 cycles/second (or 60 Hz), the armature should rotate at the speed of 3,600 rpm. If the system is formed by four poles (N1, S1, N2, S2) the armature should rotate at the speed of 1,800 rpm.

Electrons of two natures can be assumed in electromagnetism, in accordance with the proposed deuterotome theory, namely: electric-electron and magnetic-electron.

Electric-electrons are the ones that make up the flow of electrons in the electric current while magnetic-electrons make up the magnetic field that lies mainly in the magnetic poles of an iron bar, for example. Both have different characteristics: electric-electrons have a small volume, high density, low plasticity and identifiable electric load (situated internally); magnetic-electrons have a great volume, low density, high plasticity and undetectable electric load and are distributed by the magnetic field (situated externally). If one assumes that magnetic-electrons lie in the electromagnet pole, where they will then be captured, reduced in volume and conducted through the magnetic tunnel; upon leaving out through the other pole they will expand again and will return to the pole they entered originally. However, should the electrons in the poles be captured by the armatures, they will continue having a small volume, will get to the collectors in the form of electric-electrons and will follow on, forming a flow of electrons or electric current.

13.2 – Photoelectric effect

Hertz observed that the electric discharge between two electrodes occurs far more easily when ultraviolet light incides on the cathode. Later on, Lenard discovered that electric discharge happened more easily because the light falling on the cathode facilitated the emission of electrons from the surface of that electrode, a phenomenon that was called **photoelectric effect**.

The photoelectric effect was demonstrated when two electrodes were set inside a glass tube, where vacuum was produced. Externally, a polarity key, a rheostat and a micro ammeter were fixed on the circuit. No electric load is applied to the system; instead, a light beam falls on the cathode in the first test and after that on the anode, in a subsequent test. The experiment can be repeated by inverting the polarities by means of the polarity control key. There will be emission of electrons from the pole or electrode that receives the light to the pole or the opposite electrode. Because they are subject to collisions, those electrons emitted by the electrode that gets the light beam (bundle of photons) do not lose kinetic energy. The kinetic energy of electrons emitted depends on the wavelength in the photon with an incident light. Blue light, for example, has a 436 nm wavelength and green light 546 nm. The scheme ahead makes it possible to analyze this effect (Fig. 99).

In short: the photoelectric effect consists of electrons emitted by the plate of an electrode that receives a light beam, where this electrode and the other are inside a glass tube where vacuum was initially installed, with no electric load.

A microammeter C is fixed in the external circuit to show the passage of electric current from an electrode to another. The kinetic energy of the electron emitted depends on the intensity of the light source and the wavelength of the light emitted. The maximum kinetic energy from the electron (Kmax) is the result of the electron charge (e^-), multiplied by the load potential (Vs), the typical values of which are a few volts given by the equation:

$$Kmax = e. Vs$$

Assuming that an electron is linked to a metal with a work function energy W, in order for the electron to be removed from the metal, an amount of energy should be applied, at least the same as that of W, which varies from metal to metal. For example:

2.28 eV for sodium (Na11), 4.70 eV for copper (Cu29), 4.73 eV for silver (Ag47), 6.35 eV for platinum (Pt78), etc. The electron more tightly linked to the metal requires a greater amount of energy to be removed. An electron is considered free when a deuterotome absorbs a photon that has released energy at least equal to W. In other words, if **h.v < W** where **h** is Planck's constant, **v** the frequency of the electromagnetic wave and **W** the work function, there is no photoelectric effect; if **h.v > W**, the electron is free and any excess of energy appears as kinetic energy. The maximum kinetic energy for each electron released from the metal surface is determined by the following equation:

$$Kmax = h.v - W$$

A photon is a quantum of electromagnetic energy, whether emitted or absorbed. The energy of a photon **E** is expressed by the equation **E = p x C**, where ρ (rho) is the momentum of a photon of energy **E**, and **C** the light speed.

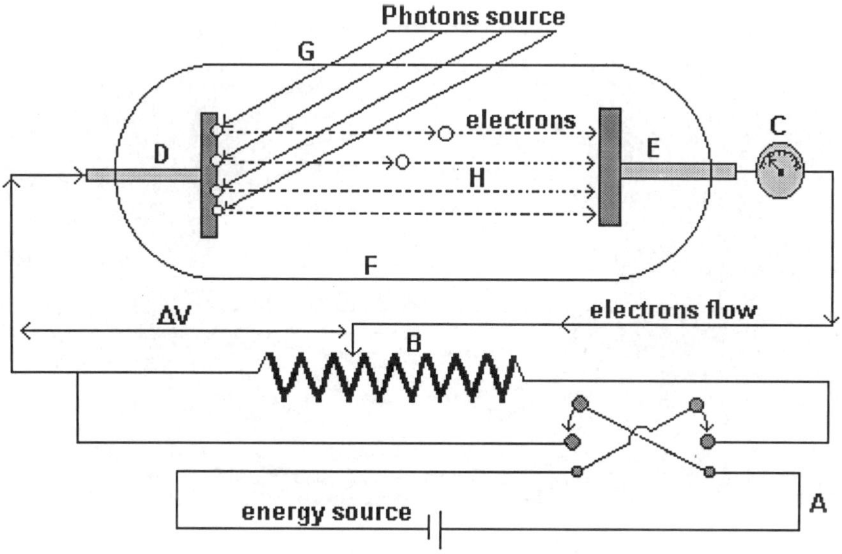

Fig. 99 - **Photoelectric effect in a vacuum tube**. **A** – Polarity control key of electrodes. **B** – Rheostat for control of the intensity of the electric current that flows from an electrode to another when there is incident light on the electrode in the tube. **C** – A microammeter to register the electric current intensity that is formed between the two electrodes. **D** – Cathode, in this case originating electrons inside the vacuum tube made of a metal plate. **E** – Anode inside the vacuum tube also made of the same metal plate. **F** – Glass tube with an internal pressure (vacuum) nearly null. **G** – Light beam (photons) falling on the cathode plate. **H** – Electrons ejected from the cathode toward the anode.

Ionization energy is the name given to the minimum amount of energy necessary to pull out an electron from a deuterotome or from an atom of light hydrogen, the latter with ionization energy of 13.6 eV. Helium (He2/4), with two electrons, in the first ionized electron has the value 24.6 eV and 54.4 eV in the second. The ionization energy is lower in alkaline-earthy metals (Li3, Na11, K19, Rb37 etc.) and higher in noble gases. Some examples can be given: Li3, 5.39 eV; Na11, 5.14 eV; K19, 4.34 eV; Rb37, 4.18 eV; Au78, 9.22 eV; Ne10, 21.56 eV and Kr36, 14.0 eV. In order to ionize the second, the third or other electrons in shells closer to the nucleus, the amount of energy necessary to remove them will be far higher.

13.3 – Magnetism

The magnetic mass is devoid of detectable electric load. A magnetized bar features two opposite poles, one for attraction and the other for repulsion. Magnetism is basically a phenomenon that occurs in relation to deuterotomes. The magnetic mass of a bar is the sum of magnetic masses in individually magnetized deuterotomes. There are two magnetic states: a **permanent state** in magnets and a **temporary state** in electromagnets. Both have to be considered in separate, for well-defined aspects exist in light of the proposed deuterotomic theory, thus making it possible to reformulate the theoretical knowledge on electromagnetic induction. The standard drawing, formed by iron fillings scattered on a magnetized iron bar, affords an initial analysis of magnetism as a physical phenomenon: in the N and S poles the fillings are laid out on a linear pattern. Concentrated close to the poles and scattered as they move away from the them, iron dusts are shaped like a fan (bidimensional); the lines in a pole that are more bent downwards appear to be linked to those on the opposite pole.

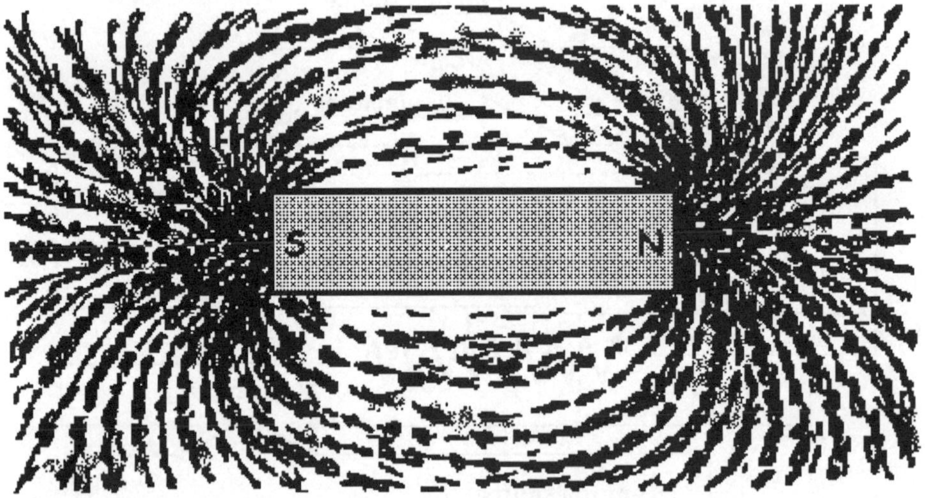

Fig. 100 - Layout assuming the granules of iron fillings in the N and S poles of a magnet. They form a negative image in the arrangement of magnetic electrons outlying the fan and free electrons in the middle of the poles.

If the granules of iron fillings are manually pushed out their lines, they will be reoriented to form their original shape, thus resuming the initial figure. Aligning the granules, where the extremity of a lengthened granule touches the extremity of another, suggests the force lines pass inside the deuterotomes. Further, each iron filling granule is magnetized and when the poles of opposite signs are attracted, a cord is formed; the filling granules kept away from one another due to the repulsion force and create an empty image among the granules which increase as they move backwards from the middle of the pole.

Thus, the figure formed by the alignment of iron filling granules is the negative in the physical phenomenon that occurs in magnetic poles. The empty spaces in the granules are the routes followed by the magnetic electrons; the spaces they travel are left empty, even though they cannot be seen or detected, unless by the magnetic field that is created in the magnet poles.

The farther the bundle of magnetic electrons from the body of the magnetized bar, the smaller the number of force lines that will act upon iron filling granules. Iron fillings recognize neither the inlet nor the outlet pole in the flow of magnetic electrons; both are identical in the negative image they perform around the bar poles. It is common knowledge that they have different functions (one intended for suction, the other for ejection) when attracted to or repelled by each other. The pole in charge of suction that attracts free electrons also attracts magnetic-electrons that are ejected by the ejector pole of the other magnet. For that reason, poles with opposite signs (N and S) are mutually attracted and repelled with identical signs (N and N; S and S).

13.3.1 – Solenoid as a generator of magnetic field

An electromagnet is a unit made up of a solenoid that includes an iron bar or a magnetizable metal alloy. The solenoid is formed by an electrically wound conductive current wire, spiral-shaped, around the bar. Inside the bar, the solenoid structure makes up a magnetic tunnel. The magnetic field is formed outside the bar. The free electrons that enter the tunnel are conducted towards the electric current that circulates in the solenoid, which starts at the source of electrons (generator) to their receptor plate. In the sketch below are the major elements that form the magnetic system of an electromagnet.

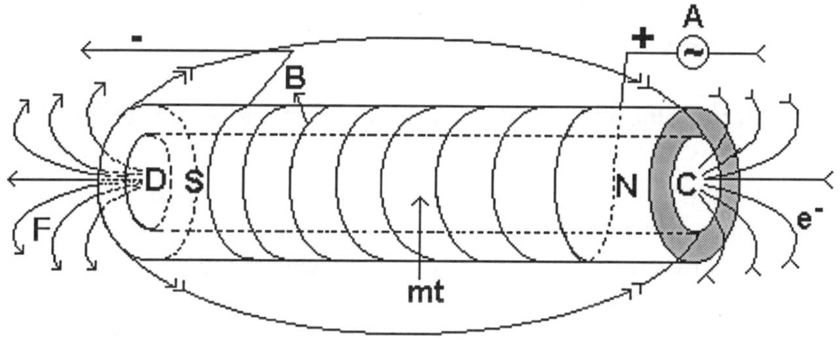

Fig. 101 – Simplified sketch of an electromagnet. **A** – Source of electrons. **B** – Solenoid, including a magnetizable metallic bar. **C** – Suction pole of the bar called N pole (north). **D** – Ejection pole of the bar termed S pole (south). **E** – Magnetic tunnel **tm** (internal). **F** – Magnetic field (external).

According to the deuterotomic theory now proposed, as far as the direction of force lines is concerned, the concept of magnetic poles is contrary to the current concept and conventional wisdom on electromagnetism. As will be seen further on, this new theory is backed up by Becquerel experiment (Antoine Henri Becquerel, 1899).

Electrons moving spirally in the conductor wire that forms the solenoid creates a magnetic tunnel in the middle of the spires, thus forcing the entrance of free electrons to the N pole (suction pole) through the interior of deuterotomes that make up the magnetizable metallic bar (soft iron or mild steel, for example), leaving through the S pole (ejection pole); both in the N and S poles, a fan-shaped image appears showing iron fillings.

The force lines outlying a pole keeps up continuity with those on the outskirts of the opposite pole, describing arch-shaped lines. The lines of central forces, that is, those leaving from the center of the pole remain straight while the ones originating from an intermediate position between the center and the periphery describe a slight bend moving outwards. No continuity occurs in the force lines in the opposite pole, though.

Because the electrons travel the solenoid in a helical motion, the free electrons that enter the magnetic tunnel heading from the N pole to the S pole should also describe the same type of motion, thus generating an **electrons whirl** inside the tunnel. When the electrons that entered the magnetic tunnel come out at the S pole, they expand in a fan-like shape. The magnetic fields in the poles fail to have a clear limit and the force of attraction or repulsion becomes gradually greater as it gets closer to the poles. When two magnetic fields located in the S or N pole approach each other, a repulsion force ensues in the aftermath. However, should a force be applied in a bid to juxtapose a pole to the other, an elastic field with a high plasticity can be observed as though two rubber balls had been pressed one against each other. This proves that the magnetic field in a pole N or S can never enter the field of another of same name; each pole has something like a field of great plasticity, the confirmation of which is provided with considerable plastic properties. This feature suggests that the electrons that had been compressed upon traveling the magnetic tunnel found their way out with great expansion force and took on a great volume (in relation to their original volume inside the magnetic tunnel), thus making up magnetic-electrons.

13.3.2 – Electric-electrons and magnetic-electrons. Essential and distinctive features

An industrial capacitor, that is, a unit built from two metal plates and a dielectric that is fixed between them, capable of storing a great amount of electrons in one of them, has no match in terms of magnetism. No insulator is known for the components forming the magnetic field or the magnetic flow that makes it possible to store magnetic-electrons. Therefore, magnetic-electrons that form a magnetic field cannot be stored with their magnetic properties preserved. Electrons departing from or entering a magnetic tunnel can be only captured by a rotating armature that balance the force lines and attracts them to the generating pole via an electrical current can only capture electrons departing from or entering a magnetic tunnel. In the magnetic tunnel of the metal bar the electrons are compressed and expand as they leave it, being compressed again when they enter the magnetic tunnel once more. Magnetic-electrons formed by expanded electrons that participate in the composition of magnetic fields, afford the plasticity feeling when two magnetic poles are approached, provided with repulsive forces (identical signs). Electrons captured by the armatures of alternated electric current generators are expanded and make up a fluid electric mass that can be stored or released in the form of electrical current in the voltage desired by means of transformers. Should the electric current that travels the solenoid be cut off, the magnetic induction then ceases as the force necessary to form the electrons whirl no longer exists to make them travel through the magnetic tunnel. An electric load can consist of electrons that have been accumulated, but magnetism will always be a flow of electrons with specific physical characteristics.

Accordingly, based on these concepts, it can be inferred that electric-electrons can be stored in a capacitor; magnetic-electrons, however can't be stored in an identical structure. Magnetic-electrons that form the magnetic field have so small a density and so great a plasticity that they can travel across empty interdeuterium spaces of materials juxtaposed to them, such as glass or wooden plates. Devoid of these two properties, they could not be compressed when traveling through the magnetic tunnel. Electrons become denser when they are captured from the magnetic poles by the armatures of electric generators, with smaller plasticity and electric load.

According, one can establish a few essential and distinctive features between these two interchangeable forms of electrons:

A) Electric-electrons have:

1) Small volume,
2) High density,
3) Small plasticity,
4) Detectable electric load,
5) Physical characteristics that enable them to be stored in electric capacitors.

B) Magnetic-electrons have:

1) Great volume,
2) Small density,
3) Great plasticity,
4) Undetectable electric load,
5) Physical characteristics that prevent them from being contained or stored in structures similar to electric capacitors.

13.3.3 – Magnetic tunnel

The tunnel that exists inside a deuterotome with a long deuterian structure is made of an actual space. Elements with a short structure such as alkaline metals (Li3, Na11, K19 etc.) and alkaline-earth metals (Be4, Mg12, Ca20 etc.) do not form tunnels and cannot be magnetized. The three metals capable of being magnetized in a sequence with Z deuterotomic numbers are: iron (Fe26), cobalt (Co27) and nickel (Ni28), all of them belonging to elements with electrons in the M-shell, series-equivalent connected. Iron is the element that has a greater binding energy per nucleon (Ben) up to uranium (U92) where the energy decreases. Nickel is the hardest of all the common metals; in other words, it cannot be scratched or perforated easily, being magnetizable below 250°C. Cobalt is the most tenacious of all metals, that is, it has great resistance to rupture when submitted to traction, flexion or torsion. Soft iron is ductile, that is, it can be stretched by a drawing mill in finer wires; gold and silver are the most ductile metals. The three magnetized elements – iron, cobalt and nickel – are resistant and hard, a property closely associated with the binding energy per nucleon, for the greatest values of that energy are found in them all.

Below is an illustration of three juxtaposed iron deuterotomes with the magnetic tunnels, each aligned and continuous to the other.

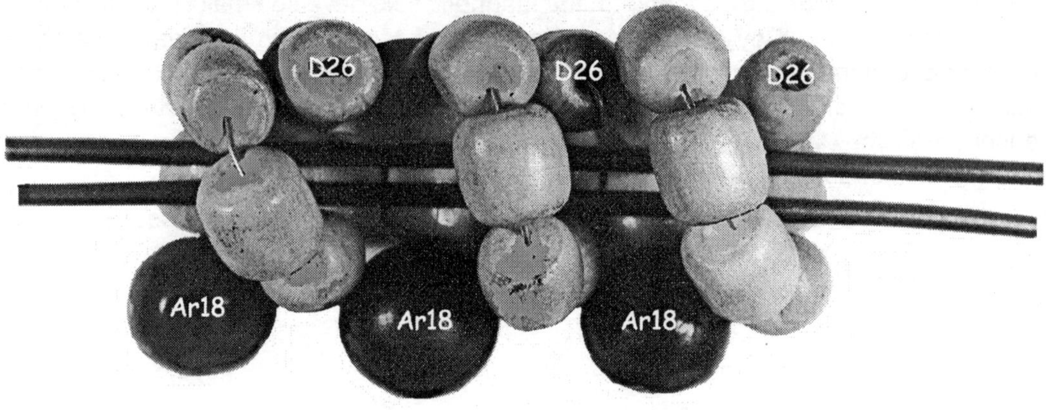

Fig. 102 – Simplified sketch of the magnetic tunnel formed by three juxtaposed iron deuterotomes (Fe26) in ring and helical shape forming a tunnel in their interior, permeable to the passage of magnetic force lines. Iron deuterotome is the magnetizable unit. The capability these deuterotomes have to juxtapose their tunnels during the lamination process affords the magnetic susceptibility. The hardness of the metal, which prevents the tunnel from flattening in, maintains its property unaffected, preventing them from being magnetizable. The same thing is true of nickel and cobalt. Noble gas argon (Ar18) is the genetic pole for these three elements; the two wires in the illustration show the permeability in the tunnels of each deuterotome. The genetic post-pole deuterons range from D19 to D26.

One other illustration also with the three iron deuterotomes, shows the magnetic tunnel occupied by force lines (compressed electrons) and the magnetic field lying externally formed by magnetic-electrons (expanded electrons).

If the north pole (N) of a magnet or electromagnet is conventionally the exit pole of force lines (left-hand rule for coils) then this conventional sign fails to comply with practice. If beta particles (β)–negative electrons–are captured by the magnet's north pole, that's because bundles of free electrons also enter through this pole. Therefore, through the magnetic north pole then enter bundles formed by beta particles (β) and free electrons (e-), which enter the magnetic tunnel. This is not an attraction resulting from an electric load, but rather due to a flow of electrons. In Becquerel experiment with radiations emanated from salts of radium (Ra88/226 and Ra88/228), the bundles of alpha particles (α2/4) emitted by radium-226 are slightly pulled by the magnetic South Pole, but not by the force of electric loads. Such phenomenon suggests that the bundles of electrons (composed of beta particles) emitted by radium-228 are attracted by the magnetic north pole (N), traveling through the magnetic tunnel and then ejected by the magnetic south pole (S). Upon leaving they exert an attraction force on the alpha particles, which have positive electric load. Gamma rays do not enter the electrons whirl because they are devoid of mass and electric load, thus not subject to the influence of either an electric or a magnetic field, for gamma rays are high frequency electromagnetic waves (photons).

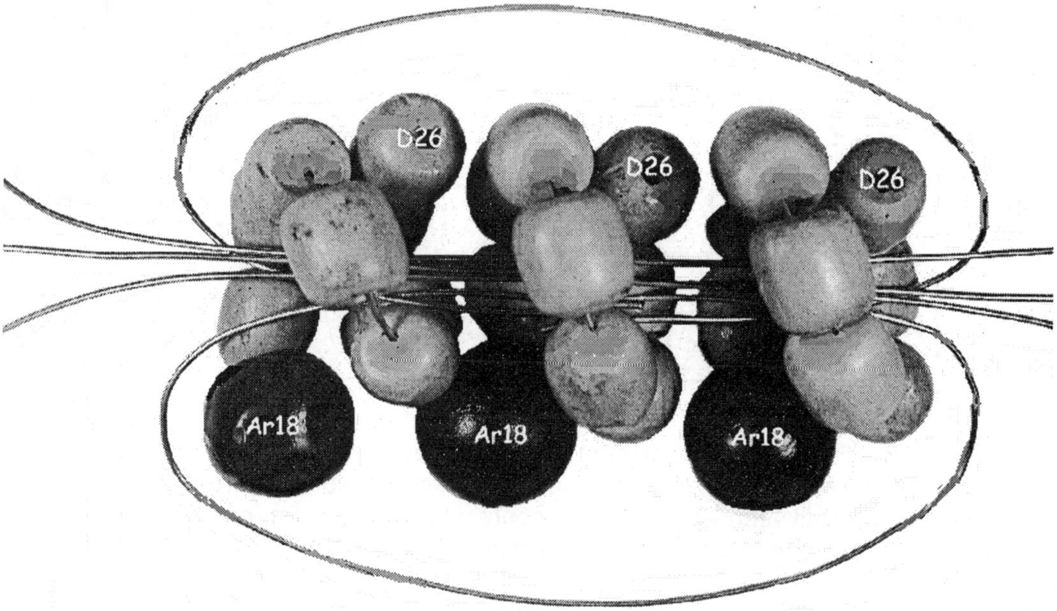

Fig. 103 - Structure of three iron deuterotomes (Fe26) with juxtaposed magnetic tunnels through where compressed electrons (electric-electrons) travel as well as and the force lines of the magnetic field, lying externally, where the expanded electrons (magnetic-electrons) are situated. See the genetic pole formed by argon (Ar18) whose volume is twice as much that of the atom of light hydrogen. See the helical shape of the deuterons genetic post-pole structure.

An electromagnet formed by a "U"-shaped iron nucleus wound by an inductor coil (solenoid) in the bar extremities demonstrates Becquerel experiment performed in 1899 by using radium salt ($RaBr_2$) in the groove of a lead block placed between the two poles (N and S) of the electromagnet.

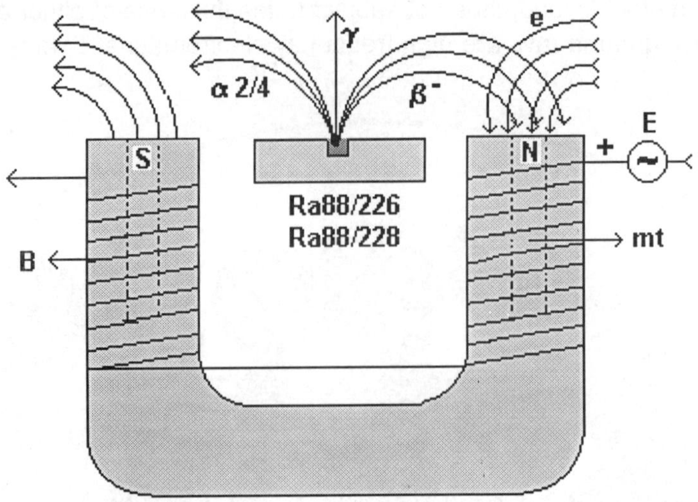

Fig. 104 - **Becquerel Experiment** – "U"-shaped electromagnet. Inductor coil **(B)** winding iron bar. **N** and **S** are the north and south magnetic poles, respectively. In the center is the plate between the poles that contain salts of radium, emitting alpha (α2/4) and beta (β) particles and gamma rays; **mt** represents the magnetic tunnel inside the bar. Free electrons (e^-) and beta particles (β^-) are attracted by the N pole (suction pole) and leave through the S pole (ejection pole). Alpha rays are slightly toward the S pole; gamma rays are not subject to any attraction. **E** is the source of electric energy that drives the electromagnet. Radium-226 only emits alpha particles while radium-228 emits both beta particles and gamma rays.

The free electrons (e^-) enter through the suction pole (N), cross the internal space of the deuterotomes (intranuclear space), accompanied by beta rays all along. The alpha rays emitted by radium-226 are not sucked in by the electrons whirl in the N pole because it can only suck up particles with negative electric load (free electrons and beta rays). Gamma rays cannot be sucked in for they are not made of matter, that is, they are devoid of both mass and electric load because they are high frequency electromagnetic waves.

The alpha particles (α2/4) are slightly slanted toward the magnetic south pole (S), not as a result of an attraction by the south pole but, most likely, because the electrons are exiting through that pole and further because their electric load is opposed to the alpha particles. The South Pole attracts them but they do not enter through it, for this pole is not the inlet for the flow of electrons, but rather, the outlet. The N pole in the electromagnet takes in beta particles because they are similar to the free electrons captured by that pole, entering the electrons whirl.

13.3.4 – Theoretical stages of magnetization

Not all elements are provided with the same capability of magnetization or magnetic susceptibility. In accordance with that property, materials are classified as diamagnetic, paramagnetic and ferromagnetic.

Diamagnetic metals are the ones that become magnetized in the opposite direction of the inductor field, thus repelling the force lines. This is the case of bismuth, for example. These metals do not possess permanent dipoles; in other words, they are deuterotomes or ions whose electronic shells are complete, like noble gases (He2, Ne10, Ar18, Kr36, Xe54). Diamagnetism is a very weak magnetic response in a system and is characterized by a negative magnetic susceptibility. The origin of diamagnetism, according to the current atomic theory, lies in the variation of the orbital angular momentum in electrons induced by the application of an electric field, to the greatness of 10^{-5}.

Paramagnetic metals are the ones that become weakly magnetized in the same direction as the inductor field; this is the case of strong electropositive elements, such as alkaline ones (Li3, Na11, K19 etc.). These metals have a small magnetization under the action of an external electric field and the magnetization occurs in the direction of the field. Nevertheless, they have a positive magnetic susceptibility to a greatness ranging from 10^{-5} to 10^{-3}. The main paramagnetic materials are as follows:

1) Metals of non-magnetic elements;

2) Materials that contain free deuterotomes and ions of transition elements of iron (from scandium, Sc21 to zinc, Zn30), lanthanides (from lanthanum, La57, to lutetium, Lu71) and actinides (from actinium, Ac89, to lawrencium, Lr103).

The origin of paramagnetism, based also on the classic atomic theory, is due to the fact that by applying a field, the conduction band of the metal is split into two sub-bands, one with an electron spin +½ and –½ respectively. This results from the fact that the energy during the magnetic momentums in each spin, in the field, is different. Of the two conduction sub-bands, the one with smaller energy has a magnetic momentum in the direction of the field and the induced magnetization holds the same direction of the field as well.

Ferromagnetic materials are strongly magnetized in the same direction of the inductor field, attracting force lines; such is the case of iron, nickel, cobalt, and other weak electropositive metals. Soft iron (that contains carbon in a content somewhat higher than 3 percent) is temporarily magnetized, while steel (iron with 0.5 to 1.7 percent of carbon) takes on a permanent magnetization under magnetic induction. Steel is more difficult to be magnetized than soft iron, but it remains magnetized even when the cause that produced the phenomenon has been interrupted, and features a magnet. The magnet attracts and magnetizes iron in contact with or close to it; if removed the magnetization is undone; if iron is kept heated up above 775°C is no longer attracted by the magnet. Should the latter be heated up above that temperature as well, it loses the magnetization and when it achieves the room temperature the magnetization is not resumed.

The magnetic induction caused by on electric current in the solenoid must at least go through three theoretical stages, which should have fractions of thousandth of second.

1st stage: forming the whirl of free electrons. When the electric current begins to trickle along the conductor wire that makes up the solenoid, the formation of a whirl of free electrons occurs inside the magnetic tunnel of the bar. As the electrons travel down the helical-shaped solenoid, the free electrons (e^-) in the environment are captured by the N pole, compressed and conducted through the tunnel (the onset of the mechanism) that ceases when the bundle of these electrons reaches the S pole. At that stage, none of the poles has a magnetic field because the magnetic-electrons have not yet been formed.

2nd stage: forming the magnetic S pole. When the compressed electrons coming from the magnetic tunnel are ejected by the S pole, they undergo a fast and progressive fan-shaped expansion process where the volume rises, the density drops and a great plasticity is acquired.

3rd stage: forming the magnetic N pole. During the time taken up by the first two stages, in no way does the N pole become magnetic, serving only as a pole for the entrance of free electrons, without a magnetic field. As the free atmospheric electrons lie very scattered or dispersed near the N pole, they are attracted by the electrons whirl that begins to concentrate near that pole, then compressed to go into the magnetic tunnel and conducted as far as the S pole. These electrons, upon leaving out through the S pole, are suddenly expanded into a fan shape, since they are now out of reach of an action by the solenoid. This expansion is truly a fan-shaped blast of these electrons, which take on the characteristics of magnetic-electrons.

As a magnetic field arises around the electric current that travels along the conductive wire-which makes up the solenoid-this field keeps on attracting magnetic electrons from the S magnetic pole toward the N pole, where they undergo a reverse process; the volume is reduced, the density is increased and the plasticity reduced. In doing so, they enter the magnetic tunnel and participate in the "pool" of electrons that give rise to the electrons whirl.

The magnetic-electrons that approach the N pole, now form the N magnetic pole with all the properties of a magnetic field-the N pole is the last to become magnetic-thus shutting off the circulation system, and is so maintained while the magnetic induction by the solenoid is held up. Through the N magnetic pole enter free electrons (e^-) and magnetic-electrons, the latter after undergoing a progressive reduction in volume and plasticity in terms of free electrons. This transformation happens gradually and quickly, as the polar surface of the iron bar is approached.

An external flow of magnetic-electrons gives rise to a magnetic field of high plasticity in both the N and S poles. These fields neither attract nor repel electric loads. The N pole attracts only free electrons from the environment. These electrons passing through the magnetic tunnel in the deuterotome or an iron bar occurs under a pressure that compacts the electrons. It is likely that the greater the number of spires and the closer to each other, the greater the compacting force. Upon leaving out through the S pole, the electrons experience a sudden decompression, go through a fan-shaped expansion process and take on the characteristics of magnetic-electrons and create a magnetic field. The magnetic-electrons themselves are likely to be bipolar and this would make it possible for the magnetic-electrons to cling to each other; otherwise they would get lost in the environment and then fail to form significant magnetic fields. Consequently, poles of identical signs are repelled because their magnetic-electrons have a free pole whose sign is identical to an iron bar pole.

The direction of the electrons whirl inside the tunnel should be the same as that of the solenoid's electric current, which goes from positive (generating source) to negative (receiving plate), that is, from the N pole to the S pole. The electrons whirl is a simple physical phenomenon. Nevertheless, if a bar of magnetizable metal is introduced in the whirl, this phenomenon becomes complex because the bundles of electrons continue to exist and need to pass through the interior of deuterotomes, a crucial requirement for tunnels to exist inside each deuterotome; further, these tunnels must be aligned to each other. The phenomenon that ensues in the iron bar is similar to that found in each deuterotome that forms it.

Iron magnetization occurs at a temperature ranging from 0°C to 750°C. Should the iron temperature lie beyond that limit, magnetization will not occur. The temperature limit for nickel is 250°C and this high temperature should set the orbital electrons lying inside the deuterotomes in an intense excitation electronic state. This will act as an **impeditive condition** for the flow of electrons whirl to pass through. Similarly, a permanent magnet submitted to a high temperature (above 750°C) loses its magnetism for that same reason: the presence of this impeditive condition, that is, the intense electronic excitation (intradeuterotomic), breaks up the magnetic system.

No insulator is known to the magnetic field, because all materials known to date are permeable to their so-called force lines naturally at room temperature and normal atmospheric pressure.

The binding energy per nucleon (Ben) that makes up the wall of deuterotomes is likely a factor determining the plastic properties of metals: **hardness** – the resistance that metals oppose to scatch; **ductility** – the property that metals possess to be stretched by a drawing mill in relatively thin strips; **malleability** – the property that can reduce metals to thin sheets by the force of a hammer or a laminator; **tenacity** – the resistance to rupture through the action of traction, flexion or torsion; **plasticity** – the greater or little ease that a metal can be permanently deformed as well as its physical state at room temperature, as is the case of mercury (Hg80)–a liquid metal at room temperature and a solid at -38,9°C. Therefore, it is obvious that the temperature is a factor that interferes in the plastic properties of metals, besides the binding energy of the nucleons. The adhesive power between deuterotomes in a liquid medium is dependent on their surface tension. When the temperature rises, the surface tension drops and the liquid evaporates. Mercury itself emits vapors at the temperature of 0°C.

13.3.5 – Solenoid

A magnetizable metal set inside the magnetic tunnel created by the solenoid, acquires magnetization while an electric load is being applied to the solenoid. As soon as the electric current ceases in the conductor wire that makes up the solenoid, the magnetic field formed in the metal bar then disappears. This phenomenon is called **temporary magnetization**. The other form, called **permanent magnetization**, is the permanent state of a magnetic field in the metal bar-the case of steel-where the electric stimulus has happened for a short period of time and created the magnetic field. Even when the electric load is shut off, the magnetic field stays put. The physical phenomenon of magnetization is the same, but the mechanism that causes it to happen is different. The new atomic theory now proposed (or deuterotomic theory) has a solid base to give an explanation as to the two forms of electromagnetism. In keeping with the current atomic theory, it is hard to formulate the same theoretical explanation as to those two forms of magnetization.

The plastic properties featured by such metals as iron (Fe26), cobalt (Co27) and nickel (Ni28) as metals provided with as high degree of hardness and tiny malleability afford their structures the capability of being unable to be deformed keeping the magnetic tunnel permeable, thus holding them receptive to magnetization. The structure of an iron deuterotome is shown in the illustration below, where deuterons D24, D25 and D26 can be seen, corresponding to the chemical pole of that element.

Platinum (Pt78), gold (Au79) and lead (Pb82) are very malleable metals, that is, they can be transformed into strips, wires, sheets, etc. by percussion, without being broken up. These metals fail to have a sustainable intradeuterotomic tunnel, breaking into a collapse, causing these spaces to become virtual; in addition, their tunnels are not aligned, and this is an impeditive condition to magnetization. Conversely, bismuth (Bi83) and white cast iron (Fe26) are very brittle.

Fig. 105 – Simplified illustration of the iron deuterotome (Fe26) structure, showing its chemical pole (D24, D25 and D26) and the tunnel within, one of the conditions that afford it magnetic susceptibility. The arrows show the direction of the force lines that travel along the tunnel.

Consequently, if there are no aligned and juxtaposed magnetic tunnels, there is no electrons whirl and the magnetic system cannot set in. The permeability available in elements is a means of relative capacity in carrying over force lines. The permeability in non-magnetizable materials such as air, wood, copper, bronze, aluminum etc. is of a Gilbert per each cm of length; the permeability of laminated iron and laminated steel is greater than that of cast iron, and reaches 1,200 to 2,600 Gilbert per each cm of length.

A magnetized steel bar or an electromagnetized soft iron bar loses its magnetization when heated up above 750°C. Even if the steel temperature should drop back to 37°C, the original property cannot be restored. Heating at so high a temperature provokes an intense excitation in intradeuterotomic electrons that cuts off the magnetic system because the flow of electrons through the tunnel is blocked up.

Fig 106 is an illustration of three iron deuterotomes (Fe26) in disarray, where the deuterotomic tunnel of one is not aligned and juxtaposed to the other. Tunnels so arranged prevent one of the basic conditions from existing so that magnetic induction can occur.

Therefore, it can be said that an element is magnetizable when it has the following characteristics:

1) Long genetic post-pole structure (beyond four deuterons),

2) High binding energy per nucleon (like iron, cobalt and nickel),

3) High degree of hardness, that is, they cannot be scratched or easily deformed by percussion,

4) Solid at room temperature,

5) Provided with a permeable intradeuterotomic tunnel, that is, not easily deformed; also during the lamination process their tunnels are aligned and juxtaposed,

6) The genetic pole is formed by noble gas argon (Ar18).

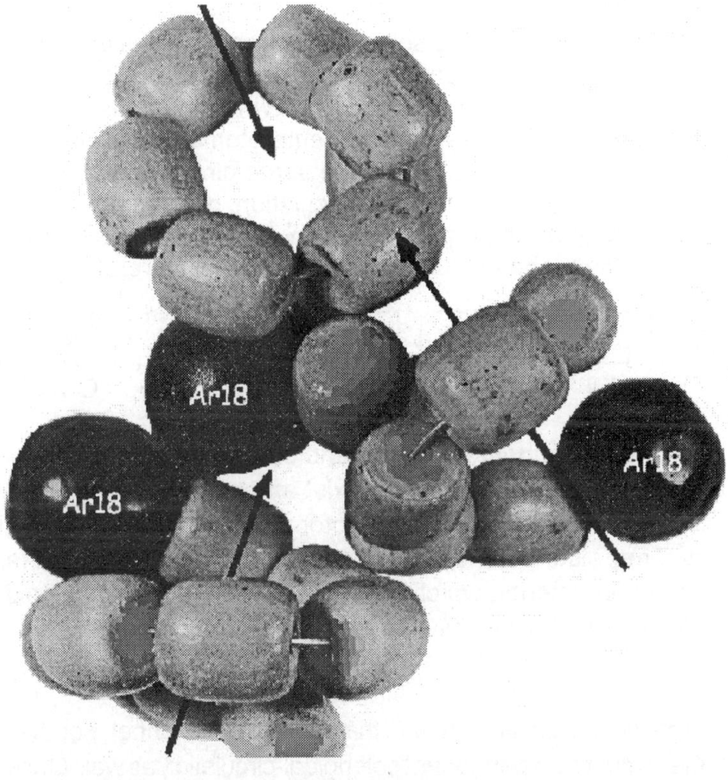

Fig. 106 - Sketch of the structures of three iron deuterotomes (Fe26). The deuterotomes are arranged in disarray and their tunnels are disordered. The alignment and juxtaposition of the tunnels are basic conditions that determine the magnetic susceptibility. The arrows show the positions and direction of the tunnels.

Such being the case, **impeditive conditions** can be found in the mechanism that triggers the magnetic system like the high temperature, for instance, 750ºC for iron and

250° for nickel. Similarly, **determinant conditions** can be also identified in this very mechanism:

1) The material should be a metal with a high degree of hardness,
2) To be provided with a deuterotomic tunnel permanently permeable,
3) To be able to align and juxtapose the deuterotomic tunnels during the lamination process at hot,
4) To have a high bending energy per nucleon (Ben).

This deuterotomic theory, composed of many other small theories, such as that of the atomic capacitor, the intradeuterotomic position of the electronic crown and the existence of a tunnel in deuterotomes with long genetic post-pole deuterium structure makes it possible to come to this conclusion. The current atomic theory cannot account for the physical magnetic phenomenon based on the principles above; according to it, the atomic nucleus neither has a well-organized structure nor a real void space, internally.

13.3.6 – Intradeuterotomic solenoidal circulation

An electric current applied to the solenoid externally, activates a magnetic system of **solenoidal-circulation** inside the tunnels of iron deuterotomes contained in a steel bar. This is so because clusters of iron deuterotomes whose tunnels are aligned have their extremities occluded, most likely by carbon deuterotomes. This obliteration prevents a long flow of electrons from heading toward the same direction and causes the electrons to turn around on the outside of the internal circulation, depending on whether the type of the solenoid is of double and continuous circulation. Carbon that is a part of the composition of steel should close the magnetic tunnel made of one or several iron deuterotomes.

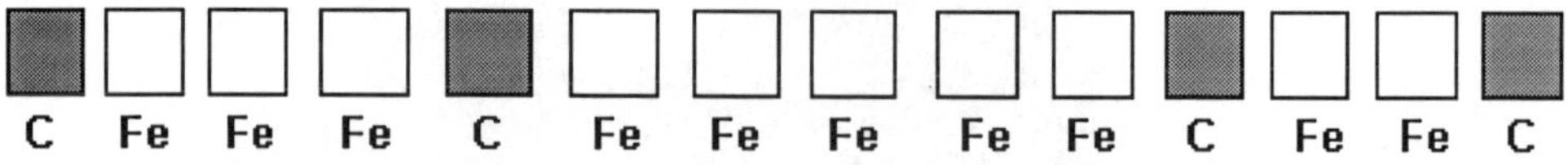

Fig. 107 – Likely arrangement of iron and carbon deuterotomes to make up steel. A great number of iron deuterotomes may exist among the carbon deuterotomes. Each segment acts as a magnet. Carbon deuterotomes limits the number of magnetizable juxtaposed units. They also limit the length of solenoidal-circulation, but fail to prevent the flow of electrons in their interior, which make up the electrons whirl inside the magnetic tunnel. The magnetic system is similar to that in an electromagnet.

The solenoidal-circulation is formed in the deuterotomic tunnel. For the magnetic field to be permanent, there must be a permanent solenoidal-circulation as well. Once the mechanism that activates the magnetic system–induced by the solenoidal-circulation–has started up, it remains active even if the external inductive cause has ceased. Movement of electrons occurs in the tunnel in a compact and central helical path, from one pole to the other. When these electrons reach the extremity of this path, they then return from the outside, following a peripheral reverse helical path with expanded spires. This prevents any significant interference in the direction of the electrons compressed inside the magnetic tunnel.

The intradeuterotomic solenoid is virtual: it does not exist inside deuterotomes. The motions performed by the electrons in the deuterotome actually take over the task of a double circulation type solenoid that is, a central solenoidal-circulation plus a peripheral one in a continuous way being the central multispiralled and the peripheral, paucispiralled. A picture of a double continuous solenoid is shown in the figure below:

Fig. 108 – Double and continuous solenoid formed by a central unit with a number of spires (multispiralled) each set beside the other with a small diameter and a peripheral unit with few spires (paucispiralled), enlarged, far from each other with a large diameter. The two units are continuous to each other, so that the electrons of solenoidal-circulation travel a long path (centrally) and a short one (peripherally), alternately.

Circulation of electrons inside a double solenoid, that is continuous and unequal, creates the condition for the permanent motion of these intradeuterotomic electrons, which cannot move out to reach the deuterotome or a group lying ahead. Thus, each deuterotome or group of them is permanently magnetized with an internal magnetic tunnel and an external magnetic field.

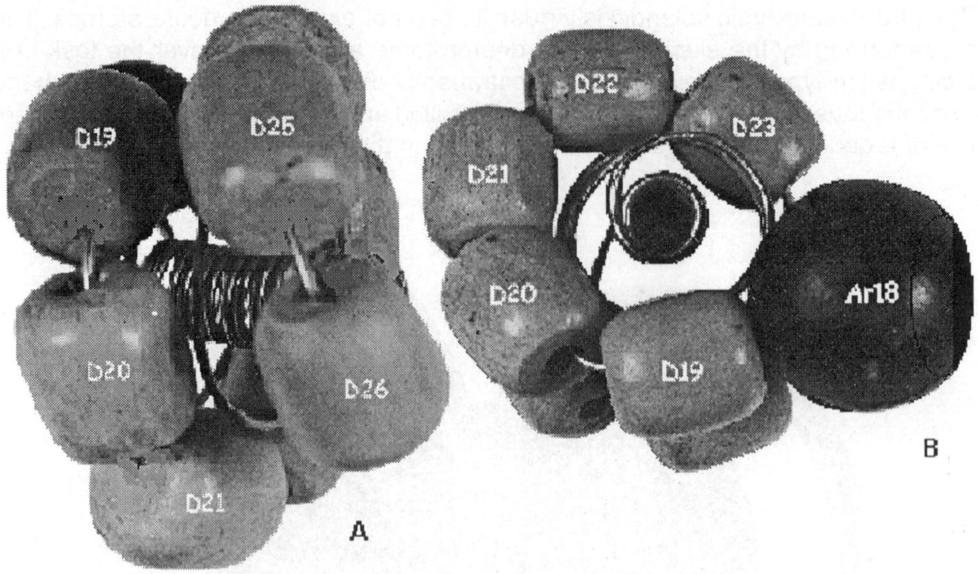

Fig. 109 – Intradeuterotomic solenoidal-circulation. Illustration of the iron deuterotome (Fe28) in its helical shape: **A** profile view; **B** top view. **A** – Illustration of a solenoidal-circulation inside the deuterotome. **B** – The sketch shows the solenoidal-circulation-continuous type-where the internal circulation is continuous to the external one. Deuterotome view from the genetic pole represented by argon (Ar18).

13.3.7 – Characteristics of magnetizable and non-magnetizable materials

Materials capable of undergoing magnetic induction possess many basic characteristics called **determinant conditions** of magnetization, which are not a phenomenon that happens at random. In order for a certain material to possess activated magnetic system, it should be provided with the following features:

1) Characteristics of a metal: no liquid or gaseous physical states,

2) Genetic post-pole structures composed of many deuterons, for example 6 to 8, as happens with iron, cobalt and nickel; a deuterotome cannot be magnetized if has a smaller number of deuterons, for example 1 to 4, as is the case of sodium, potassium, calcium, aluminum, cesium, rubidium, barium etc.,

3) Series-equivalent bindings (from K-shell to O-shell); materials possessing parallel equivalent bindings are not magnetizable (rare earth, platinum, mercury, gold, lead etc.),

4) Great binding energy per nucleon, like iron, cobalt and nickel, which afford a great adhesive force between protons and neutrons, thus forming a deuterotome with a high degree of hardness, that is, not easily deformed by pressure or percussion. Materials that are malleable or that can be laminated by pressure or percussion such as lead, platinum and gold, for example, are not magnetizable. Hard materials make it impossible for the deuterotomic tunnel to be deformed (nickel, for example), while soft metals do otherwise (lead and gold, for example),

5) Capability to align and juxtapose their deuterotomic tunnels during the hot lamination process (soft iron or steel); the alignment process occurs while the metal is being worked on between its liquid state (fusion) and the solid state (solidification) to form a bar, a sheet, a square bar-iron, a wire or a sphere.

Overall, iron cannot be said to be a magnetizable metal. Soft iron is magnetizable, but cast iron is not. The iron encountered in iron pyrites (ferrous sulphide, FeS) is not attracted by a magnetic field. The intradeuterotomic tunnel must be real, permanent, permeable, undeformable and aligned with others so as to form a long tunnel, where the electrons can make up an electrons whirl.

Below is a scheme showing the basic composition of permanently magnetizable deuterotomes.

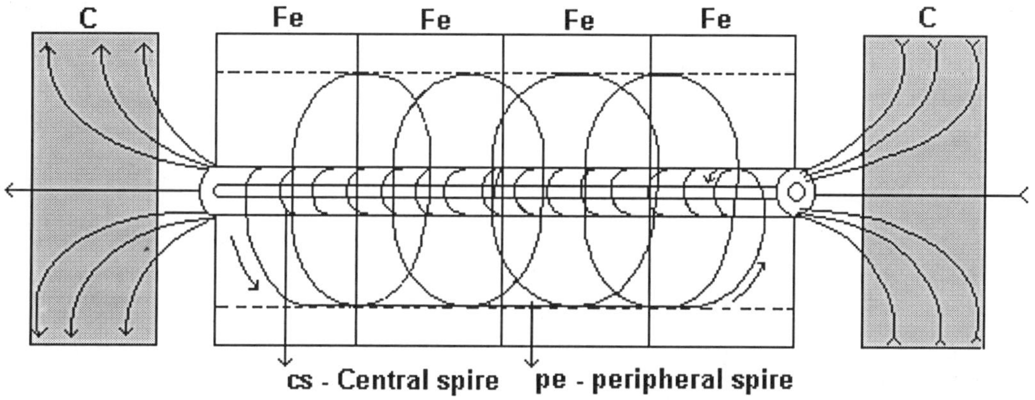

Fig. 110 –Simplified scheme of solenoidal-circulation (or solenoidal electrocirculation) inside iron (Fe) deuterotomes showing the respective tunnels aligned. Carbon (C) deuterotomes occlude the extremities of a group of aligned tunnels, thus curtailing the solenoidal-circulation, but allowing free electrons to enter through the N pole and exit through the S pole. **Cs** is the central spires and **Pe** the peripheral ones.

The figure represents a virtual solenoid and shows multiple central spires and few peripheral ones, indicating the path followed by electrons excited by the magnetic induction of an actual solenoid, situated externally. The electrons that leave through the S pole are expanded in a fan-like shape, describing an arch-shaped course and return to the N pole, entering the circulation again, thus forming the magnetic system. The electrons that travel across the tunnel are compressed, being compressed upon leaving through the S pole. The thermal electronic excitation above 750ºC in the magnet disorganizes and breaks the solenoidal electrocirculation and the magnetic field disappears. The magnetic system cannot be reset even if the initial temperature is back to its original condition.

The central solenoidal electrocirculation is a strong inductor of magnetism, while the peripheral one has a negligible effect due to the small number of spires in each unit. Once this circulation starts up, the magnetic system sets in permanently.

A possible cause for a steel bar to be permanently magnetized is the high gap energy in the carbon deuterotome, equal to 5 eV. Carbon deuterotomes occlude the extremities of the magnetic tunnel in iron deuterotomes, preventing the exit of electric-electrons that take part in the solenoidal electrocirculation. The electrons of valence band (Vb) in carbon fail to have enough energy to cut across strips of high-energy gap (Eg) and, in doing so, reach the conductivity band (Cb), thus obstructing the flow of an electric current (flow of electrons). The magnetic-electrons that make up the magnetic field flow freely across carbon and any other matter.

CHAPTER 14

Some insights into the Deuterotome Theory

The deuterotomic theory now proposed holds a few points in line with the classic atomic theory. The atomic theory arose from a myriad of questions, such as the Periodic Table of Elements published by Mendeléeff in 1781, which contains 63 well-known elements and leaves plenty of room for over 20 elements to be accommodated, unknown until then. Prior to reaching that stage in knowledge, a great deal of researchers made progress little by little. In 1829 Doebereiner observed that elements that were chemically similar formed triads. An example of this is **lithium-sodium-potassium**, whose mass difference is 16 (Li3/7, Na11/23, K19/39); other triads, like **calcium-strontium-barium**, have a mass difference of 48 (Ca20/40, Sr38/88, Ba56/136), in addition to others such as **sulfur-selenium-tellurium**, **chlorine-bromine-iodine**, provided with identical masses.

According to the so-called **law of the octaves** formulated by Newlands in 1865, elements were established in an upward scale as per their atomic masses, where an entire set of seven elements was followed by an eighth element whose properties are similar to that in the element occupying the same position in the previous set. A story is told about a wise man attending a lecture on the law of the octaves who asked Newlands whether he could not come up with a similar law with the elements classified in alphabetical order. This question highlights that difficulties existed at that time so that issues associated with number, mass, and periodicity of elements could be perfectly assimilated. Not that this is a ground rule, but where there is no knowledge, there is no comprehension.

The Periodic Table of Elements by Mendeléeff failed to leave room for noble gases; in other words, even though empty spaces were left for many unknown elements, no room was available for the noble gases discovered by Lord Rayleigh and Sir William Ramsay between 1894 and 1898.

The electronic theory brought to light new progresses made on knowledgeable areas, above all the works carried out by Moseley (Henry G. J. Moseley, 1887-1915, English, died on World War I battlefield), who defined that the property of elements lies in the periodical function in their atomic numbers; the works he performed on the X-ray spectrum unveiled the first relationship between the Chemical Periodic Table and Atomic Physics.

The discovery of radioactivity and X rays are closely associated, the latter discovered by Roentgen in 1895. Because X rays bring about phosphorescence in certain substances, for instance, zinc sulfide, barium platinocyanide etc., in light of the principle of reciprocity it was then inferred that certain phosphorescent substances were capable of releasing X rays. In 1896 Henri Becquerel insolated a number of substances (the material was left in the sun), to convey them phosphorescence. In doing so, he found out that double sulfate of uranium and potassium, was able to impress photographic plates and then came to the conclusion that this substance, when insolated, produced X rays. On one occasion, because it was a cloud day making it, thus impossible to insolate uranium salt, Becquerel put it away in a drawer along with a photographic plate. When the plate was revealed he noticed that it came out impressed, obviously an evidence that there was one other form of radiation other than X ray that was capable of producing the same effect due to a property in that salt which he called **radioactivity**. Even when the Curies were trying to isolate highly radioactive elements in certain pitchblendes, these turned out to be three or four times more radioactive than the oxide of pure uranium, those radioactive properties were not yet very well-known; next the Curies fractioned the pitchblende and came across highly radioactive salts of barium and bismuth. Pierre Curie discovered polonium associated with bismuth and Marie Curie discovered radium associated with barium in 1898; in 1900 she succeeded in isolating metallic radium.

The atomic theory was developed based on what was known then about the physical and chemical properties of elements and molecules. The findings on electricity and magnetism, backed by those theories on electronics, increased considerably the number of discoveries and their actual applications. Valves diodes and later transistors revealed that the basis for theories is supported by unquestionable facts. Also unquestionable facts about the atomic theory in the fields of radioactivity, nuclear fission and nuclear fusion are the atomic explosions of plutonium bombs (nuclear fission) and hydrogen bombs (thermonuclear fusion).

The quantum theory unfolded a wide range of findings on electrons and photons that helped account for and set up energetic values in the field of atomic physics, impracticable without them. On April 30, 1897, Joseph John Thomson announced the discovery of a new and luminous particle (electron), working in the Cavendish Laboratory in the University of Cambridge in Great Britain. Without that knowledge or the discovery of electron the quantum theory would not exist because the quantum is a package of energy released by the electron when it strikes a target wall or when it travels from an orbit onto another with lower energy. That theory, in terms of electrons, was instrumental in broadly expanding knowledge as far as the electronic crown of atoms is concerned. Nevertheless, the findings on the atomic nucleus, basically formed by protons and neutrons, are still in the dark. The quantum theory can only reach the interior of the nucleus as regards the emission of gamma rays. The nuclear fission of uranium-235, due to the action of slow neutrons and emission of alpha particles (originating from the nucleus) for example, fails to have plausible enough theoretical physical grounds. The current atomic theory contains a wide range of knowledge acquired as to the performance of electrons in orbital zones; as regards the nucleus, though, they are relatively scarce; the nuclear structure is impenetrable to be fully deciphered and assumptions cannot be formulated, let alone a theory, when one intends to find terra firma in its structure, despite the theory of the magic numbers.

The deuterotomic theory was based on the assumption that space is full of units called lentons, devoid of mass or well-defined forms and devoid of energy when considered separately. If there were no lentons in space, it could not exist. When two or more lentons approach, touch and join each other, they store up electric load and become form of energy. Lentons form plates and make up electrons and protons; an electron is formed by n x 1 lenton and a proton by n x 1,836 lentons. The lentonians plates are separated from each other by a space called lentonian interstice, more often found in protons. A single lenton or clusters of them can be housed in both, making up small energy packages just like a photon that is a quantum of energy. Thus, a small lenton package can possibly be a quantum of energy; this similarity is possible between them.

The lenton theory as to the formation of protons and electrons is individually in line with this new deuterotomic theory, that is, unrivalled by the atomic theory in force.

The deuterotomic theory mingles with the current atomic theory or complies with it when it brings to light, heavy and super-heavy hydrogen atoms.

The deuterotomic theory one dichotomize of the current atomic theory from helium (He2/4) deuterotome onwards based on the atomic capacitor theory, where the capacitive unit consists of two plates, namely the proton (receptor of electron and photons) and the electron (polar complementarity) and by a dielectric, i.e., the neutron; this dielectric can be single in deuterium and double in tritium. Thus, deuterium is the single mono-dielectric atomic capacitor and tritium the bi-dielectric one.

Because deuterium is a single capacitor, helium (He2/4) turns out to be an element formed by two capacitors or two deuterons. From this concept, helium is no longer an atom, but rather a **deuterotome** and this designation has adopted for all the elements from helium itself to uranium and trans-uranium.

The "Table of possible ways for the synthesis of deuterotomes", besides the synthesis mechanisms "neutron capture and emission of beta particles", includes the "capture of alpha particles", the "neutrons capture", the proton-proton cycles and the carbon cycle. Apart from this, a possible way for the synthesis of heavy, radioactive deuterotomes is available by "hyper-capture of neutrons" from thallium deuterotomes (Tl81/202, 203 and 204), crucial to establish the phylogenetic code in the alpha appendix of heavy and radioactive deuterotomes.

According to the deuterotomic theory developed, the extraordinary physical volume-reduction phenomenon in noble gases–for they are electrically neutral–makes it possible to include them as the genetic pole of deuterotomes which guide the synthesis that add number and mass to deuterotomes due to their properties via "capture of alpha particles" and "neutrons capture". The genetic pole is a deuterotomic capacitor with so higher a capacitance as the number of its protons, that is to say, with capacitive plates where the plates are series-equivalent connected, which happens to noble gases from helium (He2/4) to xenon (Xe54).

The parallel-equivalent bindings in post-xenon deuteriums up to radon (Rn86) form a cluster composed of 32 electrons near the genetic pole, termed "electronic furnace", which, for their turn, do not go through the volume-reduction phenomenon. This cluster of electrons remains permanently excited, releasing electrons that strike the deuterotome's chemical pole to which they belong, thus causing the rupture and the disjunction of an alpha unit (α2/4) and its consequent emission, called alpha particle. The electron thrown against the deuterotome wall can also be subject to diffraction in keeping with the degree of incidence and return to the "furnace" immediately or after experiencing other diffractions. The electric tension inside a heavy deuterotome should be high, for its electrons belong to orbital zones of high energy levels (Rn86 (Xe) $6s^2$ $5d^{10}$ $4f^{14}$ $6p^6$).

The spiraled or helical settings of long genetic post-pole deuterotomes allows the existence of an internal, real space, inside the deuterotome, the more permeable the higher the degree of hardness of the element (metal), this hardness resulting from the higher binding energy per nucleon (ben).

The series-equivalent or parallel equivalent bindings demonstrate that the orbital electrons in the chemical poles can be kept free of the capacitive bindings and be changed into free valences. Because these electrons lie in the opposite extremity of the genetic pole and are linked to chemical reactions, that extremity was termed **chemical pole**.

The chemical pole of radioactive deuterotomes in the families of radium, actinium, uranium-235 and thorium, the alpha-appendix is also available which keeps a specific phylogenetic code in each radioactive family. The code in each alpha-appendix holds the position of those neutrons that will release beta particles (β) in a preset sequence, a fact that prevents the emission of alpha particles ($\alpha 2/4$) if the alpha-unit is not specifically prepared to do so, even if there are electrons of the "furnace" hitting the alpha-appendix.

The existence of an intradeuterotomic tunnel has led to the magnetic system theory, which accounts for the formation of magnetic fields, the temporary–the electromagnet– and the permanent–the magnet. With respect to electricity, in what it refers to the theory of valence and conduction bands, as well as the energy gap (in the interval zone), there is no divergence. As regards magnetism, though, the deuterotomic theory comes up with a new theory as to how the magnetic field is formed. Here appears the concept of a **magnetic tunnel** that conducts an **electrons whirl** whose direction is identical to the electric current traveling across the solenoid, the poles equally designated by N and S. Here it is admitted that N is the entrance pole and S the exit pole, thus contradicting the left-hand rule. This inversion is based on a tangible proof, the experiment by Becquerel, who demonstrated that beta rays (β) are captured through the N pole, as proposed by this deuterotome theory, during the first stage or phase of magnetization. It also creates the concept of electric-electrons, with small volume, high density, low plasticity and detectable electric load, that make up the electrons whirl inside the magnetic tunnel and the magnetic-electrons, with great volume, small density, high plasticity and undetectable electric load, that form the magnetic field. It creates the concept of intradeuterotomic solenoidal electrocirculation in magnets with permanent magnetization.

The theory on the emission of gamma rays through the capture of an electron-K by a proton situated intranuclear is in line with this deuterotomic theory with no divergence. According to the former, an excited nuclear state and one of electronic excitement, is created movement of electrons on orbital levels with higher energy, which occupy the shells with lower level and start to emit photons. The emission of Auger electrons complies with the deuterotomic theory as well.

The new theory proposed herein also confers a new aspect upon the physical phenomenon of radioactivity or the radioactive property in heavy elements belonging to the uranium (U92) and thorium (Th90) families. The production of both **continuous** and **characteristic** X rays in elements (also termed **discrete**) is a physical phenomenon whose production mechanism in no way differs from the new deuterotomic theory, since continuous X rays are produced by emission of photons when an electron emitted by cathode of the Crookes' tube hits a target wall and because the characteristic X-ray results from an electron passing from an orbital zone with high energy level into another with lower level to occupy the vacancy left by the electron-K captured by the nuclear proton. The physical phenomenon mentioned above could take place in electrons that comprise noble gases forming genetic poles of elements, even when their volumes are reduced. Because the production of discrete X rays is dependent on a nuclear phenomenon, the capture of an electron-K by a nuclear proton, and an electronic phenomenon, an electron of higher energy level moves toward a lower one to occupy the vacancy cascade which starts from electron-K likewise as this phenomenon can occur inside the deuterotome of a noble gas, the deuterotomic theory now proposed also complies with the current atomic theory. The electrons that make up the electronic crown of noble gases are "armored electrons" with respect to the carriage system of electric current in keeping with the bands theory, but not with regard to the other physical phenomena such as the capture of an electron-K by a nuclear proton, the electrons that move from an orbital zone with higher energy level toward one with lower level, the emission or absorption of photons, the emission of Auger electrons, etc.

The electrons in the "electronic furnace" in heavy and radioactive deuterotomes, upon hitting the target wall composed of deuterons, can emit photons which can remain inside the nucleus or, rather, escape through the chemical pole; the phosphorescence color appearing in the emitting deuterotome is dependent on the kinetic energy in the photon produced: **E = h x v** where the photon wavelength determines the color, within the visible spectrum. Thus, a photon with a wavelength of 436 nm (420/470) gives off a bluish phosphorescence while a wavelength of 546 nm (505/554) gives off a greenish one. Reddish phosphorescence can be seen with a wavelength of 700 nm (670/760). The hues in the colors depend on accurate wavelengths, because a certain color meets a band within variable lengths and visible light varies in a range from 400 to 700 nm, that is, from violet to red.

Accordingly, the proposed deuterotomic theory is also quantum–based as it enlarges the quantitative notion of matter–allowing a possible identity between lentons (that fill up space and most possibly form electrons and protons) and photons (that are ejected from the interior of an electron when it strikes the target or when it moves from an upper energy level to a lower one in the atom's orbital zones). What's more, based on the theory on the "electronic furnace", electrons and photons alike are set inside heavy and radioactive deuterotomes. Besides, the theory takes for granted that protons can possibly release photons as well, should they go through intense nuclear excitation, because lenton packages can be found in the interstices of lentonian plates that constitute a proton (as they constitute an electron). The difference as to the likelyhood of photons (or lenton packages) being released between the electron and the proton lies in the fact that the electron is mobile, agile, fast, and capable of becoming a true electron-projectile, just like the electron released by the cathode of Crookes' tube that hits the target and loses photon or energy quantum, while the proton is a fixed unit (when it composes the deuterons in the deuterotome walls) or a mobile unit when it makes up the atom of hydrogen or its ions; under no circumstances, however, is it provided with mobility and speed like those of an electron.

The deuterotomic theory is quantum–based from lentons to protons and electrons, to the radioactive composition of heavy elements, which emit gamma rays. No basic concept is altered in physics or well-known physical laws. If anything, it brings to mind that the neutron is a unit that can be laid out in the column of noble gases as shown in the Periodic Table of Elements, right above helium, for it is a neutral unit just like noble gases and never yield its atom status. If the proton of an atom of light hydrogen could capture its electron, thus giving rise to a neutron, a gaseous mass of neutrons would ensue as noble gases do, that is, they would be inert units. It is common knowledge, however, that the capture of an electron by a proton can only take place inside an atom. Because the neutron is not reactive, albeit capable of being captured as is uranium-238, should it change into uranium-239, an unlimited number of possibilities would arise.

To make a broader comparative analysis between the proposed deuterotomic theory and the classic atomic theory, would right now be out of keeping with this book, as the identities, the divergences and the increments existing between them all should be thoroughly underlined.

Bibliography

1–COTTON, F. A.; WILKINSON, G. and GAUS, P. L. *Basic Inorganic Chemistry.* Third Edition. John Wiley & Sounds, Inc. N. Y., New York, USA, 1995.

2–KNOLL, Glenn F. *Radiation Detection and Measurement.* Second Edition. John Wiley & Sounds, Inc. N. Y., New York, USA, 1989.

3–KRANE, Kenneth. *Modern Physics.* John Wiley & Sounds, Inc. 605 Third Avenue, N. Y., New York, 10158, 1982.

4–MELLO-LEITÃO, Cândido. *Glossário Biológico.* Cia. Editora Nacional, Rio de Janeiro, RJ, Brasil, Second Edition, 1946.

5–MELLOR, J. W. *Química Inorgânica Moderna.* Translation: by Alcides Caldas (original title: Mellor's Modern Inorganic Chemistry). Editora Globo S/A., Porto Alegre, RS, Brasil, Second Edition, 1952.

6–MURRAY, Raymond L. *Introduction to Nuclear Engineering.* Prentice-Hall Inc. Englewood Cliffs, N. J., USA. Sixth printing, March 1957.

7–REZENDE, Sérgio M. *A física de materiais e dispositivos eletrônicos.* Editora da Universidade Federal de Pernambuco, PE, Brasil, 1996.

8–PERSON, Elizabeth F. *Introdução à Física Nuclear.* Editora da Universidade de São Paulo (USP) e Editora McGraw-Hill do Brasil Ltda., 1978.

9–U. S. NAVY, NAVAL BUREAU OF PERSONNEL, TRAINING PUBLICATIONS DIVISION – *Curso completo de Eletricidade Básica.* Translation: given by the Centro de Instrução "Almirante Wandenkolk", Ministério da Marinha, RJ, Brasil (original title: Basic Electricity). HEMUS Livraria Editora Ltda.